Energy Code Essentials

Based on the 2018 International Energy Conservation Code®

International Code Council
Stephen Kanipe
Shaunna Mozingo
William H. (Bill) Hudson

ENERGY CODE ESSENTIALS:
Based on the 2018 International Energy Conservation Code®

Stephen Kanipe
Shaunna Mozingo
William H. (Bill) Hudson

International Code Council

ICC Staff:

Executive Vice President and Director of Business Development:
Mark A. Johnson

Senior Vice President, Business and Product Development:
Hamid Naderi

Vice President and Technical Director, Products and Services:
Doug Thornburg

Vice President, Publishing and Multimedia:
Margaret M. Leddin

Senior Marketing Specialist:
Dianna Hallmark

ISBN:	978-1-60983-703-7
Project Head:	Steve Van Note
Publications Manager:	Mary Lou Luif
Project Editor:	Mary Lou Luif
Production Technician:	Beverly Ledbetter
Cover Design:	Julia Lange

Errata on various ICC publications may be available at www.iccsafe.org/errata.

First Printing: July 2018

PRINTED IN THE USA

T022959

Preface

Construction practice, building design and material development have changed little over many centuries. People mostly lived in temperate parts of the world, and cold living spaces were managed simply by layering up with heavy clothing and lighting a fire. Hot temperatures were often avoided by changing living habits to be less active during the day and seeking shade and a breeze. Some cultures developed portable shelters to avoid the heat. In all cases and throughout most of history, people built *something* to support their need to live in a safe and comfortable place. Building practice developed over time to suit the local climate and make use of local materials, with the common goal of keeping out dust, rain, threatening animals and bugs and letting in light and fresh air.

Fire safety has long been a concern as well. As more buildings were constructed in denser city patterns and more people gathered in larger buildings, fire issues had to be addressed. When loss of life and massive property damage become intolerable, the need to regulate construction is satisfied by building codes. The regulatory environment aims to fulfill people's expectation that when they go to work in a building or to an event in a large concert hall, they will breathe clean air and get out safely. The regulations are crafted for the understood need for safety of living and sleeping in the comfort of our homes. The codes adopted by governmental agencies develop over time to refine structural practice so that buildings withstand the forces of nature to protect the building occupants and provide safe, clean water and air.

But buildings need to do more than just shelter people and business. As the structural fire resistance and exiting, plumbing, and ventilation systems of buildings became more reliable and the cost of energy increased, building and design professionals began to focus on reducing energy consumption. Regulating energy use in buildings is a relatively new concept in the design and code enforcement industry. Nationally recognized building safety regulations were published in the early 1900s. The first national energy code was published in the mid-1980s. Code development related to building safety principles has been organized for over a hundred years. Designers, tradespeople and administrators have had a lot of time to test, try and teach modern safe building practice. By comparison, the focus on energy efficiency is new to many in the building design and construction trades. The *International Energy Conservation Code*® (IECC®) developed quickly into a complex document, and the learning curve was steep. The complexity is necessary as innovation in buildings and building systems increases. As new materials, methods and equipment are introduced, the code provisions change to keep up with advancements. The regulations may overwhelm the homeowner, designer or builder unfamiliar with the energy code. Sorting through the complex and detailed provisions can be intimidating.

Energy Code Essentials captures the provisions that regulate energy use in commercial and residential building construction. It is written to provide a readable and user-friendly overview of the IECC, explaining regulations in clear, noncode language. Understanding energy code provisions is essential to the application of the IECC to any building design; thus, this text is illustrated to further simplify and communicate essential concepts. The text is presented and organized in a user-friendly style with an emphasis on technical accuracy and clear, understandable language. This book is directed to readers familiar with basic construction, architectural and building system principles but with a limited knowledge of energy code requirements and provisions.

Anyone involved in the design, construction or regulation of building construction can learn from this book. Homeowners, people in nearly any building trade and those in building design will gain a basic understanding of the principles, provisions and applications of the technical content of the energy code. The reader will be able to use this book to more fully explore the most common residential and commercial energy efficiency code provisions.

The content of *Energy Code Essentials* is organized into commercial and residential building provisions and discusses the administrative regulations that a code official in the building department will use to enforce the energy regulations. The climate zone map and how climate zones affect requirements for different parts of the country are explained, as well as how to use the map. Examples of work that requires a permit and, just as important, work that does not need a permit are identified. Coverage is also provided of the level of detail and information to be included in the building plan documents that are prepared for permit submittal. Requirements for what construction and equipment need to be inspected and when to call for the inspection are explained. The provisions for basement and wall insulation and discussion about the choices that may best suit a particular condition are introduced in easy-to-follow code references. Requirements for windows, doors and skylights and their selection and installation in accordance with the IECC are covered. The provisions to control air leakage are included in the specific requirements section of the book (Part IV). Efficient mechanical system specification and installation provisions are included, along with what to do to specify and install a compliant hot water system. Discussion also includes the provisions for electric power and regulations applicable to exterior and interior building lighting.

Even with this extensive coverage, this book is not intended to explain all of the provisions of the commercial and residential energy code or all of the acceptable materials and methods of construction. It focuses on the most common and used provisions applicable to many conditions in residential and commercial construction. This is not to say that the information not covered is any less important or less valuable to the reader. This book should be used with the *2018 International Energy Conservation Code*, which should be referenced for more detail.

Reasonable application of the code provisions is supported by a basic understanding of the scope and intent of both the IECC and the other International Codes® (I-Codes®). This book and the IECC reference other codes and standards, as such a basic understanding of interrelated concepts and provisions provided in this book is necessary.

Energy Code Essentials contains full-color photos and illustrations to help the reader visualize and understand the application of the code requirements. Practical examples, simplified tables and highlights of particularly useful information help clarify the basic requirements of the code and help the reader determine compliance. References to the applicable sections of the 2018 IECC are included to easily locate the applicable code section for more detail and exact code language. A glossary of energy code terms clarifies the meaning of technical terms. This is the third edition of this publication, updated for the 2018 IECC.

ABOUT THE INTERNATIONAL ENERGY CONSERVATION CODE

The IECC is a comprehensive model code that regulates minimum energy-efficient provisions for new buildings and additions and alterations to existing buildings. There are two separate sets of provisions. The commercial and residential regulations each apply to heating, air-conditioning, ventilation and lighting systems. Administrative provisions and definitions specific to each commercial and residential set of regulations are also included. The IECC integrates easy-to-understand prescriptive provisions for compliance as well as performance criteria that make possible the use of new materials, new equipment and new building designs.

The IECC is one of the codes in the family of the International Codes published by the International Code Council (ICC). All of these codes are maintained and updated through an open code development process and are available internationally for adoption by the governing authority to provide consistent and enforceable regulations for the built environment.

ACKNOWLEDGMENTS

Energy Code Essentials is the result of many hours of research, code language analysis and collaborative effort. The authors are grateful for the valuable assistance and contributions of Jay A. Woodward, ICC Senior Staff Architect. Jay was the ICC staff secretary to the IECC Code Development Committee during the fast-moving early years of sweeping energy code changes. His unwavering commitment to accuracy and to those actively involved in the tedious code hearing process is much appreciated. The knowledge and experience of the ICC staff were instrumental to its development, and Jay contributed to the accuracy and quality of this product. Hamid Naderi, PE, Senior Vice President of Business and Product Development at the ICC, came up with the concept of this book and provided the initial direction. Dr. Joseph Lstiburek, Shaunna Mozingo and the Colorado Chapter Energy Code Change Committee continues to contribute to the ever-growing body of knowledge relating building and energy codes to high-performance building practice and are highly appreciated for their important work. The authors would like to thank Pam Cole, Research Scientist in the Energy Policy & Economics Group at the Pacific Northwest National Laboratory, Robby Schwartz of Energy Logic and Howard Wigg of the Hawaii State Energy Office for their review of this edition. Lastly, a special thanks to Joseph Kistner, Special Projects Intern in the Aspen Building Department, for the LED research, developing graphics, managing files and organization for the residential chapters of this edition.

ABOUT THE AUTHORS

William H. (Bill) Hudson, CBO, MCP
Building Official
Oak Brook, Illinois

William H. (Bill) Hudson, CBO, MCP, is currently Building Official for the Village of Oak Brook, IL. Bill has a broad base of experience; as facilities manager, general contractor, inspector, plan reviewer, building official and code instructor. In addition to his regular duties that involve the day-to-day application of the codes, Bill has had the privilege of participating in ICC code training since 2003 as subject matter expert, technical reviewer, author and instructor. As an instructor, Bill has had the opportunity to teach both in the United States and internationally. In addition to training activities, Bill is a past-president of Suburban Building Officials Conference (SBOC), and currently serves on four U.L. Standards Technical Panels.

Shaunna Mozingo
Building Plans Analyst, Inspector and Code Educator/Consultant for Colorado Code Consulting, LLC

In addition to being a plans analyst and inspector for many jurisdictions, Shaunna manages the energy code contracts for CCC, conducting research and providing training to code officials, builders, energy raters and designers, both locally and across the country. Shaunna also has served as a building inspector and competition juror for the U.S. Department of Energy's (DOE's) Solar Decathlon Competitions and served as the inspector for the China Solar Decathlon Competition. She has been awarded the DOE's Jeffrey A. Johnson Award for excellence in the advancement of building energy codes and performance, IMT and ICC's Standard Bearer's Award for her efforts in compliance with energy codes as well as the Southwest Energy Efficiency Project's (SWEEP's) Award for Leadership in Energy Codes. Shaunna is a past president of the Colorado Chapter of ICC and serves on the Chapter's Education, Legislative, Code Development and Membership committees. She was appointed to the International Code Council's Commercial Energy Code

Development Committee and Code Correlation Committee for the 2018 and 2021 code cycles.

Andrew Klein, PE, CEM
Principal, A S Klein Engineering, PLLC
Pasco, Washington

Andrew Klein is the principal of A S Klein Engineering, PLLC, a professional engineering firm specializing in energy management and high-hazard code consulting services. Prior to forming his own firm, Mr. Klein worked for six years as a process engineer responsible for the design validation of a $12 billion nuclear waste treatment plant.

Mr. Klein is actively involved in the code development process of the International Code Council (ICC) and serves on two National Fire Protection Association (NFPA) committees and one International Association of Plumbing and Mechanical Officials (IAPMO) committee. Mr. Klein earned his bachelor of science degree in Chemical Engineering at the University of Delaware and his master of science degree in Engineering and Technology Management from Washington State University. He is a licensed Professional Chemical Engineer in Washington State, a Certified Energy Manager (CEM) through the Association of Energy Engineers (AEE) and an ICC-certified Energy Code Specialist.

Stephen Kanipe, CBO, LEED AP
Chief Building Official
Aspen, Colorado

Stephen Kanipe is the Chief Building Official of Aspen, Colorado. He is the author of the residential chapters of this edition. My Kanipe has 29 years of experience in code administration, plan review and field inspection. His career started with, and continues to be supported by, the Aspen Community Development Department. He was appointed to his current position in 1995. Mr. Kanipe was selected by the Board of Directors of the International Conference of Building Officials to participate in the International Energy Conservation Code Development Committee and served for five years, including two as chair of that committee. In May 2009, Stephen was appointed to the ICC's Sustainable Technology Building Committee to help develop the *International Green Construction Code®* (IgCC®) and continues to serve on the Energy/Water IgCC Code Development Committee. He is a member of the Colorado Chapter of the International Code Council Past Presidents Committee. Mr. Kanipe was instrumental in developing the Aspen and Pitkin County Renewable Energy Mitigation Program, which was recognized in May 2007 by Harvard University's John F. Kennedy School of Government as a finalist in the Ash Institute Innovations in Government Award. He received an associate of science degree in Architectural Technology in 1988 from Columbus State College in Columbus, Ohio. Mr. Kanipe is a Certified Building Official and LEED Accredited Professional.

ABOUT THE INTERNATIONAL CODE COUNCIL

The International Code Council is a member-focused association dedicated to developing model codes and standards used in the design, build and compliance process to construct safe, sustainable, affordable and resilient structures.

Most U.S. communities and many global markets choose the International Codes. ICC Evaluation Service (ICC-ES) is the industry leader in performing technical evaluations for code compliance, fostering safe and sustainable design and construction.

Governmental Affairs Office:
500 New Jersey Avenue, NW, 6th Floor
Washington, DC 20001-2070

Regional Offices:
Eastern Regional Office: (Birmingham)
Central Regional Office: (Chicago)
Western Regional Office: (Los Angeles)

Phone: 888-ICC-SAFE (888-422-7233)
Website: www.iccsafe.org

CONTENTS

PART

I

Introduction to Energy and Building Codes

Chapter 1: Energy and Building Code Perspectives

Chapter 2: Legal Aspects, Code Adoption and Code Official Authority

CHAPTER 1

Energy and Building Code Perspectives

© istockphoto/mustafa deliormanli

INTRODUCTION

Building codes are the various sets of regulations related to the construction, alteration, maintenance and use of buildings and structures. The codes offer a common and familiar guide for adoption by governmental agencies to approve building plans and inspect construction. There are 15 separate International Codes® (I-Codes®), covering topics including structural design for earthquake, hurricane and tornado forces; provisions for fire and life safety; provisions for energy conservation; guidelines on systems for heating, cooling, plumbing and electrical utilities and guidelines on efficient material, land and resource use (Figure 1-1). These codes serve primarily to protect the safety and welfare on building occupants and the public, and to reduce the negative impact of the built environment. The *International Energy Conservation Code®* (IECC®) provides design guidelines for the effective use and conservation of energy in commercial and residential buildings. The IECC regulates building components in exterior walls such as insulation, windows and doors; the efficiencies of electrical equipment such as lighting, motors and transformers and the performance of heating, cooling and service water heating equipment. The IECC references companion I-Codes for elements of construction outside the scope of the IECC, such as mechanical, plumbing and electrical system installations. This section briefly discusses the history of energy codes and the scope and influence of some of the companion codes and their relationship to the IECC.

FIGURE 1-1 The 2018 *International Codes.*®

HISTORY OF ENERGY CODES

The development of energy codes and standards began in an orderly fashion that continues to be used today. The American Society of Heating, Refrigerating, and Air-Conditioning Engineers (ASHRAE) was founded in 1894 and published Standard 90 in 1975. This standard, titled *Energy Conservation in New Building Design,* was developed in 3 years in accordance with the new National Bureau of Standards (NBS) *Design and Evaluation Criteria for Energy Conservation in New Buildings*. Standard 90, now ANSI/ASHRAE/IESNA 90.1-2013 (Figure 1-2), is recognized in the 2018 IECC Section C401.2 as a path to compliance for commercial energy efficiency.

FIGURE 1-2 *International Energy Conservation Code,* 2018 Edition, including ASHRAE 90.1, 2016 Edition.

Model code organizations followed much the same path developing the *Model Code for Energy Conservation* in 1981, and then the *Model Energy Code* (MEC) in 1983. The MEC (Figure 1-3) was prepared and maintained by the Council of American Building Officials (CABO). This is a significant collaboration in that the MEC brought the then three national code groups—the International Conference of Building Officials (ICBO), the Southern Building Code Congress International (SBCCI) and the Building Officials and Code Administrators International (BOCA), along with the National Conference of States on Building Codes and Standards—together for a common cause: to write an energy code for building officials and the building industry. The U.S. Department of Energy (DOE) provided the contract to create the 1983 MEC with content based on ASHRAE Standard 90-1980. The document was organized in the common code format and still exists in the structure and organization of the 2018 IECC.

FIGURE 1-3 The Model Energy Codes by the Council of American Building Officials.

THE CASE FOR BUILDING AND ENERGY CODES

Modern construction codes are continually developed and improved to help make our built environment safe, healthy and energy efficient. Code history reaches back more than 3,500 years to Hammurabi, whose code created the intent of building regulation based on fairness; construction is to be sound and buildings are to last—consumers should get what they pay for. Fire-resistant construction and improved sanitation principles were implemented as Rome rebuilt after its AD 64 fire. This marked a shift in focus from protecting the building owner and builder to primarily risk reduction.

Perhaps as a result of recalling European tragedies, building codes first appeared in the United States in 1625. These early codes were concerned with fire safety and roof coverings. For example, in 1630, Boston prohibited chimneys made from wood. Thatch roofs were outlawed at about the same time. George Washington suggested height and area limitations on wood-frame buildings in the District of Columbia. In 1788, the first formal building code in the United States was written in Old Salem, now Winston-Salem, North Carolina.

The cost of nonregulated construction is an issue of national importance. In the summer of 2011, the U.S. Congress considered the Safe Building Code Incentive Act (H.R. 2069). Testimony during the hearings stated "Adopting and enforcing newer codes can reduce losses (from natural disasters) by 40 percent or more."

The cost of energy-inefficient buildings is also a matter of national importance. The DOE's *Building Energy Codes Resource Guide,* published in June 2011, states "Building energy codes are estimated to produce a financial benefit to owners of nearly 2 billion dollars annually by 2015, rising to over 15 billion dollars annually by 2030." In this context, energy costs are a slow-moving disaster. The realized savings are risk reduction and lower utility costs for building owners and occupants.

You Should Know

cdpACCESS™

- cdpACCESS is ICC's new cloud-based system for the code development process (cdp).
- It was developed to increase participation in the code development process.
- cdpACCESS allows users to create, collaborate, review, submit and vote (if eligible) on code change proposals and public comments.
- After the Committee Action Hearings, ICC members can view and vote on motions for those code changes that received an assembly motion.
- After the Public Comment Hearings, ICC will post the Online Governmental Consensus Vote. The proposals and hearing testimony are available to be viewed by everyone; ICC Governmental Member Voting Representatives and Honorary Members will be able to vote.

CODE DEVELOPMENT

Construction technology, methods, materials, equipment and processes are constantly evolving. Accordingly, the International Code Council (ICC) model codes (I-Codes) are systematically revised at periodic intervals to keep up with technical changes and to address the improved understanding of hazards. The I-Codes are revised and updated through an open process that invites participation by all stakeholders and affected parties. This code development process can involve exhaustive research, review, discussion and debate of the potential changes.

A code change begins with the submittal of a proposal to add, revise or delete a code provision. Any interested individual or group (other than ICC staff) may submit a code change proposal and participate in the proceedings in which it and all other proposals are considered.

Proposed code changes are submitted online through the ICC cdpACCESS™ (code development process access) website. Following the publication and distribution of the proposals, an open public hearing, the Committee Action Hearing (CAH), is held before a committee comprising representatives from across the construction industry and government, including code officials, contractors, builders, architects, engineers, industry professionals and others with expertise related to the applicable code, or portion thereof, under consideration. This open debate and broad participation before the committee reflects a consensus of the construction community and those impacted by building and fire codes in the decision-making process. The committee may approve, modify or disapprove the code change proposal.

If there is a successful motion from the floor regarding the committee's action, then the assembly floor motion will become available for online balloting after the CAH concludes. All members of ICC can participate in the Online Assembly Floor-Motion Votes. If the assembly floor motion receives a majority vote, then the motion becomes a Public Comment to be heard at the Public Comment Hearing. Once the first stage of online voting is completed, the results of the CAH and the subsequent online voting are compiled and become available as the Report of Committee Action Hearing (ROCAH). Following the availability of the ROCAH, anyone may submit a public comment proposing to modify or overturn the hearing results.

The next public hearing is the Public Comment Hearing, in which the merits of code change proposals that received public comment are debated. Any interested party may offer testimony; however, only ICC Governmental Member Voting Representatives (designated public safety officials of a government jurisdiction responsible for administering and enforcing codes) and ICC Honorary Members are permitted to cast votes at the Public Comment Hearing. Public safety officials have no vested financial interest in the outcome, and they legitimately represent the public interest. Votes are cast electronically at the Public Comment Hearing, and the results are then made available online at cdpACCESS. In the Online Governmental Consensus Vote, again only the ICC Governmental Member Voting Representatives and ICC Honorary Members are permitted to vote. The voting at the Public Comment Hearing and the Online Governmental Consensus Vote will determine the revisions to the next edition of the code.

A new edition of the code is published every 3 years. Each I-Code is developed during a single 12-month cycle during the 3-year period. There are three separate code change cycles within each 3-year period, with individual codes assigned to one of the three cycles. This important process ensures that the I-Codes reflect the latest technical advances and address the concerns of those throughout the industry in a fair and equitable manner (Figure 1-4).

Code Essentials

International Code Development cycle:

1. Other than ICC staff, anyone can submit a code change proposal using cdpACCESS™.
2. Proposals are published and distributed.
3. Open public hearings are held before the committee at the Committee Action Hearing.
4. Results are published and distributed.
5. Anyone can submit public comments on hearing results using cdpACCESS™.
6. Public comments are published and distributed.
7. An open Public Comment Hearing is held.
8. Votes are cast by ICC Governmental representatives and Honorary members present at the Public Comment Hearing.
9. ICC Government members and Honorary members can vote in the Online Governmental Consensus Vote via cdpACCESS™.
10. A new edition of the code is published.

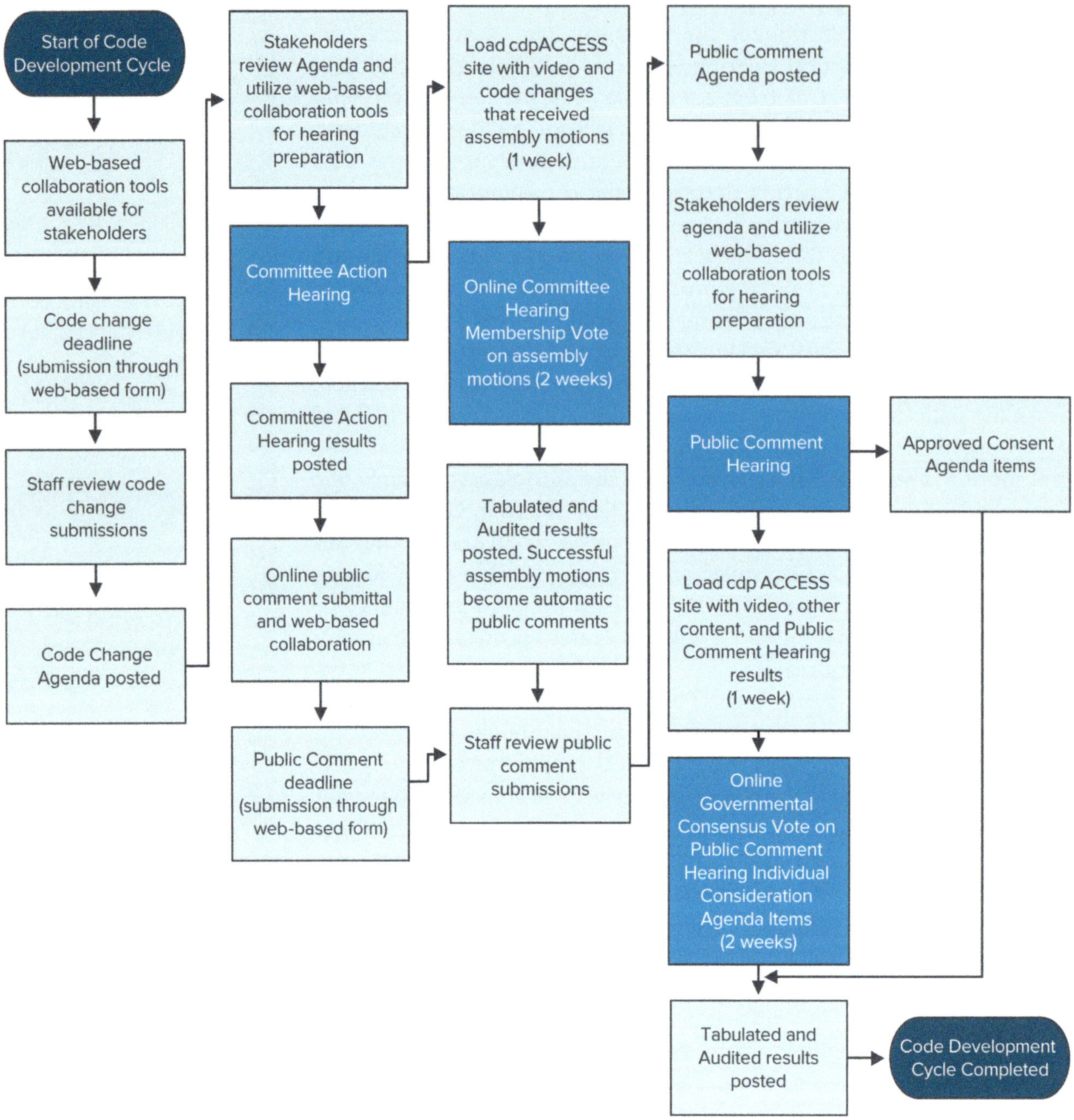

FIGURE 1-4 Code change flowchart.

THE BUILDING CODES: SCOPE AND LIMITATIONS

A number of features are common to all of the I-Codes. Each code begins by stating its scope of application. The scope establishes the range of buildings, uses, construction, equipment and systems to which the code applies. A purpose statement follows the scope and includes the intent to provide minimum standards to protect the health, safety and welfare of the public. Subsequent sections place limitations on the application of the code. For example, each code permits the continued legal use of existing buildings if the buildings do not create hazards to the occupants or property and meet certain minimum acceptable standards for health and sanitation; such buildings do not need to be brought into compliance with the current codes. In the case of an addition to an existing building, for instance, only the addition need comply provided it does not cause an unsafe condition in the existing structure. Each I-Code also references other codes and standards for use under specific circumstances. For example, the IECC references the *International Mechanical Code*® (IMC®) for installation of heating, cooling and ventilation equipment. The IMC also has requirements for public health and safety, such as to remove automobile exhaust from a parking garage. These provisions are beyond the scope of the IECC. The efficiency and performance of the equipment is within the scope of the IECC. Together, the 15 I-Codes provide a comprehensive body of regulations to be adopted and enforced to provide citizens a minimum acceptable level of protection of life and property from fire and other hazards and adequate light, ventilation and egress. Finally, the appendices of each code are not in effect unless they are specifically adopted by the local authority jurisdiction.

You Should Know

Codes are minimum requirements to safeguard public safety, health and welfare and to regulate the effective use and conservation of energy in buildings. ●

International Building Code (IBC)

The provisions of the *International Building Code*® (IBC®) apply to the construction, alteration, maintenance, use and occupancy of all buildings and structures (Figure 1-5), except those covered by the *International Residential Code*® (IRC®). In addition to structural components and systems, the IBC includes provisions for a safe means of egress, accessibility for persons with disabilities, fire resistance, fire protection systems, weather resistance, and finishes and interior environments. These regulations are typically related to the use and occupancy of the building. That is, the IBC assesses relative risks or hazards based on the functions within the building and controls design accordingly. Provisions regulating the building's size, means of egress, fire-resistive elements and fire protection systems vary significantly among different types of buildings, for example, sports arenas, hospitals, schools (Figure 1-6), apartments and office buildings.

FIGURE 1-5 *International Building Code*, 2018 Edition.

FIGURE 1-6 A school building regulated by the IBC.

FIGURE 1-7 A single-family home regulated by the IRC.

International Residential Code (IRC)

The *International Residential Code®* (IRC®) regulates the construction of one- and two-family dwellings and townhome structures. These particular buildings are not included in the scope of the IBC. The IRC provisions include all of the regulations for building structural elements, fire and life safety, mechanical, fuel gas, plumbing, electrical and energy in one document. This creates a stand-alone code for residential construction that provides strong, stable and sanitary homes that conserve energy while still offering adequate lighting, comfort conditioning and ventilation. The IRC is one source of regulation for homeowners, homebuilders and jurisdictions. It provides many prescriptive provisions for construction of homes (Figure 1-7).

FIGURE 1-8 *International Mechanical Code,* 2018 Edition.

International Mechanical Code (IMC)

The provisions of the *International Mechanical Code®* (IMC®) generally apply to the installation, alteration, use and maintenance of permanent mechanical systems utilized for comfort heating (Figure 1-8), cooling and ventilation (HVAC) and other mechanical processes within buildings (Figure 1-9).

FIGURE 1-9 Hotel mechanical equipment installation regulated by the IMC.

International Fuel Gas Code (IFGC)

The *International Fuel Gas Code®* (IFGC®) regulates the installation of natural gas and LP-gas piping systems, fuel gas utilization equipment, gaseous hydrogen systems and related accessories (Figure 1-10). The fuel-gas piping system extends from the utility company point of delivery to the equipment shutoff valves. Code coverage includes pipe sizing and other design considerations, approved materials, installation, testing, inspection, operation and maintenance. The equipment installation requirements include combustion and ventilation air, approved venting, and connection to the piping system.

FIGURE 1-10 *International Fuel Gas Code,* 2018 Edition.

International Plumbing Code (IPC)

The provisions of the *International Plumbing Code®* (IPC®) generally apply to the installation, alteration, use and maintenance of plumbing systems (Figure 1-11). The IPC includes the material and installation requirements for water supply and distribution, plumbing fixtures, drain waste and vent piping, and storm drainage systems.

FIGURE 1-11 *International Plumbing Code,* 2018 Edition.

FIGURE 1-8 *International Green Construction Code,* 2018 Edition.

International Green Construction Code (IgCC)

The *International Green Construction Code®* (IgCC®) is the newest member of the I-Code family. It has been developed to safeguard the environment, public health and the general welfare of people, places and resources affected by development and buildings (Figure 1-12). The provisions intend to reduce the negative impacts and increase the positive impacts (Figure 1-13) of the built environment on the natural environment and building occupants.

FIGURE 1-13 High-rise residential building with installed wind-powered electric generation.

FIGURE 1-14 *International Energy Conservation Code,* 2018 Edition.

International Energy Conservation Code (IECC)

The *International Energy Conservation Code®* (IECC®) intent statement is curiously different from those of the other codes and does not speak specifically to life safety and property protection (Figure 1-14). In fact, the statement points out that the IECC is not intended to reduce any safety, health or environmental requirements in other applicable codes or ordinances. The intent of the IECC is to "regulate the design and construction of buildings for the use and conservation of energy over the life of each building." Further, the code is intended to provide flexibility to permit the use of innovative approaches and techniques to achieve this objective, like the other I-Codes, through alternative materials and methods of construction, design or insulating systems.

It is this divergence from life safety and property protection that makes the energy code unique in the family of I-Codes. Many involved in the design, construction and inspection of buildings recognize that now is the time to focus our understanding on the energy code. The 2018 IECC meets the need for a design and construction document in common code format to address energy-efficient building enclosures and mechanical, lighting and power systems.

The IECC, as well as *Energy Code Essentials: Based on the 2018 International Energy Conservation Code,* acknowledge distinct differences in requirements for commercial and residential buildings. Commercial and residential requirements are organized in separate provisions of the 2018 IECC. The requirements for permits and inspections, climate zones and code administration are almost the same for commercial and residential construction, but insulation, lighting and mechanical requirements greatly differ. Even the definitions used to establish the common vocabulary for commercial and residential energy requirements vary. The commercial provisions are located in IECC Chapters 3 [CE], 4 [CE] and 5 [CE] and are included in Parts II and III of this book. IECC Chapters 4 [RE] and 5 [RE] contain the residential provisions. These requirements are addressed in Part V of this book.

CHAPTER 2

Legal Aspects, Code Adoption and Code Official Authority

Building codes establish minimum regulations intended to benefit citizens while living in, working in or visiting structures in the built environment. Most model code provisions focus on the health and safety of the occupants. The intent of the *International Energy Conservation Code*® (IECC®) is to reduce building energy use and increase the comfort of our homes and offices. Just like the "life safety" codes—building, plumbing and mechanical—the IECC must be legally adopted to become law. This section describes the adoption process and explores local amendments required for effective administration and enforcement of the energy code provisions.

CODE ADOPTION

The Constitution of the United States of America grants states jurisdiction over regulation of building construction. Each state operates by specific statutes legislating authority regarding the adoption of building codes. Some states regulate building construction by appointing a specific board or agency. The statutes of many states allow the local county, city or recognized governmental organization to adopt a code to regulate building construction. The IECC is designed to be adopted by reference by ordinance. The adoption must include the specific local information in the bracketed locations in the sample figure.

Changes to the International Codes® (I-Codes®) are made every 3 years, and the updated versions, especially in the field of building energy use, recognize new materials, developing technology and efficient and effective methods of construction. It is important for homeowners, designers and contractors to be familiar with all the codes and amendments adopted and enforced in the jurisdiction in which their project is being designed and built. This action is a necessary first step in the building permit process.

Amending the IECC

Model codes provide the advantage of consistency and uniformity in construction code adoption in a jurisdiction and across jurisdictional boundaries. Uniformity is good for material and equipment manufacturers, designers, builders and code officials in charge of administering and enforcing building regulations. It is a fact that every model code adoption requires amendments to reflect specific local conditions. Some states and jurisdictions incorporate many amendments in the adopting ordinance and others very few changes to the code text as published.

The energy code may be one of the codes a jurisdiction chooses to amend. This is necessary to acknowledge unique local climate and weather conditions that do not agree with the broader climate zones published in the IECC. Amendments are also driven by politics, local building practice and tradition, and required compatibility with other state or local laws. Any amendment included in the adopting ordinance is part of the law and must be recognized by all users regarding the energy code provisions.

Codes and standards

Codes and standards are sometimes mentioned in the same sentence and may be thought to be very similar documents. Code users should know the difference. A code is a body of laws that helps a city, state or country establish rules of compliance. The Code of Hammurabi, for example, is a well-known ancient set of laws, and rule number 229 speaks directly to building regulation: *If a builder builds a house for someone, and does not construct it properly, and the house which he built falls in and kills its owner, then that builder shall be put to death.* This is a performance code. The house should not fall in, but provisions are not offered to

You Should Know

Chapter 6 of the 2018 IECC provides the listing of referenced standards. Applicable standards of organizations such as ASTM International, ASHRAE and CSA are included. ●

guide the builder in exactly how to prevent this occurrence. Prescriptive code language is a provision that details exact requirements for compliance, such as Section R402.2 of the IECC, "Specific Insulation Requirements," which addresses ceilings, walls, slabs and other building components with information on exactly what the builder shall do to comply with the law. When "shall" is used in a building code provision, the regulation is mandatory.

Standards are developed to establish a target of performance. Materials and methods are measured and tested to determine performance and compliance with the applicable standards. This issue is so important that referenced standards get their own code sections and chapter in the building codes. It is generally easy to connect code language and standard applicability. Chapter 6 in both the commercial and residential code provisions lists the referenced standard with the publication date, edition year, title and code section referencing it (Figure 2-1). The specific standard edition provided to demonstrate code compliance must agree with the standard in Chapter 6. The specific code provision applies when there is a conflict between the code and the standard.

Code Essentials

Jurisdictional amendments become part of the code provisions and are required to be made available to permit applicants. Know the amendments in the jurisdiction having authority to review plans and issue the permit. The following is an example of an amended IECC definition.

> The definition of "Residential building" in Chapter 2 [CE] and Chapter 2 [RE] is hereby amended to read as follows:
>
> **RESIDENTIAL BUILDING.** Means a detached one-family or two-family dwelling or any building that is 3 stories or less in height above grade that contains multiple dwelling units, in which the occupants reside on a primarily permanent basis, such as a townhouse, a row house, an apartment house, a convent, a monastery, a rectory, a fraternity or sorority house, a dormitory, and a rooming house; provided, however, that when applied to a building located within the boundaries of a municipality having a population of 1,000,000 or more, the term "residential building" means a building containing one or more dwelling units, not exceeding 4 stories above grade, where occupants are primarily permanent. ●

Federal law

Energy efficiency is near the top of the national agenda. The potential economic benefits realized in building-related manufacturing, product development and employment for a skilled workforce are incorporated in the American Reinvestment and Recovery Act of 2009 (ARRA). This legislation provided funds to states to educate code officials, builders, designers, engineers and public officials regarding the positive economic impacts of adopting the IECC.

The ARRA provisions reference the 2009 IECC provisions for residential buildings and ANSI/ASHRAE/IESNA Standard 90.1-2007 for commercial buildings, or equivalent codes, to establish effective and efficient building performance regarding energy use. The ARRA, signed on February 14, 2009, includes provisions adopting the most recent edition of the energy code. This is important because the ARRA required that states demonstrate 90-percent compliance by 2017 as a condition of receiving the funds. As the 2018 IECC has been determined by the U.S. Department of Energy (DOE) as more efficient for both residential and commercial construction, adoption of the most recent edition of the IECC moves states closer to compliance with the ARRA. As a result of the ARRA, the federal government recognizes two documents, the 2009 IECC and ASHRAE 90.1-2007, as benchmarks for energy efficiency.

You Should Know

Several federal agencies are actively involved in the building safety and energy efficiency arena, such as the Department of Energy (DOE), Department of Housing and Urban Development (HUD), Department of Justice (DOJ), National Institute of Standards and Technology (NIST) and Federal Emergency Management Agency (FEMA). ●

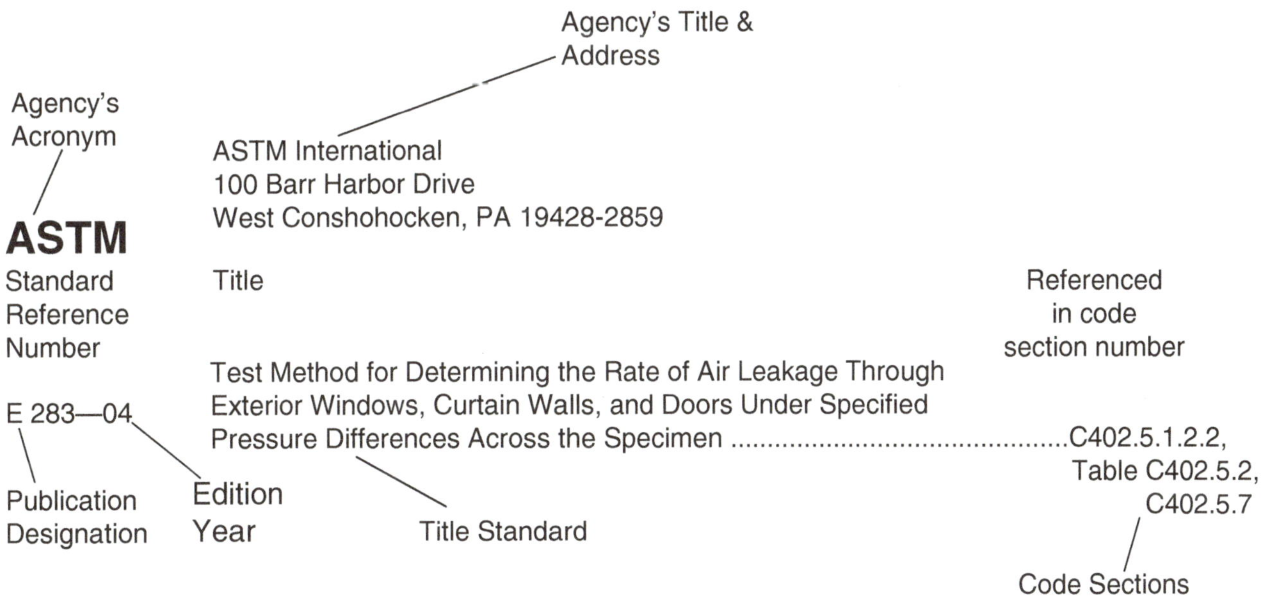

FIGURE 2-1 Referenced standard example.

AUTHORITY

The IECC grants the code official authority to enforce the requirements in the energy code. When it is adopted, the code becomes law and, like any other law, must be enforced. The code official establishes a fee schedule and cannot issue a permit until all fees are paid. Work that requires a permit may not start before the permit is issued. The code official has the authority to stop work that has not been approved and fine the person if the order is ignored and the work continues.

The code official also establishes a board of appeals to hear problems permit holders may have with decisions or orders made concerning the application and interpretation of the energy code. This board hears such an appeal, makes a decision and delivers it in writing to the appellant and code official. Thus, the code grants broad authority to the code official but creates the board of appeals to ensure the authority does not go unchecked.

DUTIES OF THE CODE OFFICIAL

The code official is charged with issuing plans reviewed for code compliance and inspecting the construction for compliance with the approved plans and the energy code. The code official may use several methods to achieve this primary function. Alternate materials, methods of construction, or design or insulating systems may be used when approved by the code official. Other energy code programs may be considered in compliance with the IECC if the code official of the jurisdiction determines that the alternate program exceeds the energy efficiency requirements of this code. **[Ref. C102.1.1]**

Plan review

The code official also is responsible for accepting plans and documentation for review and performs this duty by requiring that information specific to energy compliance be included in the building plans. The documents must clearly show the where, what and how of the proposed work. If the details, sections, insulation, windows, mechanical design, ductwork and lighting plan do not sufficiently include the information needed for energy code review, the code official may send the applicant back to the drawing board. The plans are submitted and reviewed prior to the start of construction so that noncompliant details or conditions can be corrected to save costs and avoid problems after a structure has been built (see Figure 2-2).

An ICC Evaluation Service (ICC-ES) report provides information required to approve the specified product (Figure 2-3). The report describes the product, details installation instructions, lists specific conditions of use and identifies a compliant product.

When the code official accepts project plans and documentation, a plan review for compliance with the IECC requirements must be performed. as well as a review of building life safety, mechanical, plumbing and electric code requirements. When the code official determines that the plans comply with all the applicable codes, the permit documents are stamped "REVIEWED FOR CODE COMPLIANCE" and issued (Figure 2-4). The code official keeps one set and another is kept on the job site in the field. These approved plans are for the builder to follow and the code official to reference during inspections (Figure 2-5). **[Ref. 103.3.1]**

Inspections

Any work or construction that requires a permit also requires inspection. The complexity of the project determines the construction process and, therefore, the inspection sequence. This is because the energy code does not allow work to continue beyond what has been inspected and approved by the code official. Specific inspections for wall, ceiling and floor insulation; windows; duct insulation and mechanical and water heating efficiency are required by the *International Building Code®* (IBC®) and reiterated in the IECC. To illustrate this, if a basement wall is approved with exterior insulation, the area must not be backfilled before the below-grade insulation inspection is called for and approved. Work that does not comply must not be covered up until approved to do so by the code official. Likewise, if a wall cavity insulation inspection fails, the

FIGURE 2-2 Plans reviewer at work.

ICC EVALUATION SERVICE

Most Widely Accepted and Trusted

ICC-ES Evaluation Report

ESR-XXXX

Issued August 2014

This report is subject to renewal August 1, 2015.

www.icc-es.org | (800) 423-6587 | (562) 699-0543 *A Subsidiary of the International Code Council®*

DIVISION: 07 00 00—THERMAL AND MOISTURE PROTECTION

Section: 07 24 00—Exterior Insulation and Finish Systems

Section: 07 24 19—Water-Drainage Exterior Insulation and Finish System

REPORT HOLDER:

SAMPLE COMPANY
1234 EAST AVENUE
ANYWHERE, CALIFORNIA 91000
(800) 123-4567
www.samplecompany.com

EVALUATION SUBJECT:

SAMPLE LCR SYSTEM, SAMPLE II LCR AND SAMPLE III LCR SYSTEMS

1.0 EVALUATION SCOPE

Compliance with the following codes:

- 2012 and 2009 *International Building Code®* (IBC)
- 2012 and 2009 *International Residential Code®* (IRC)

Properties evaluated:

PROPERTY	IBC CHAPTER	IRC CHAPTER
Weather resistance	14	R7
Structural – transverse wind load resistance	16	R6
Fire-resistance-rated construction	7	R3
Exterior insulation and finish systems (EIFS)	14	R7
Surface burning characteristics	26	R3

2.0 USES

The Sample LCR system, Sample II LCR system and Sample III LCR system are exterior insulation and finish systems (EIFS) complying with IBC Section 1408 and IRC Section R703.9. The systems comply as an EIFS with drainage in accordance with IBC Section 1408.4.1 and IRC Section R703.9.

The systems may be used in fire-resistance-rated, Type V construction when installed in accordance with Section 4.5 of this report.

3.0 DESCRIPTION

3.1 System Components:

See Table 1.

3.2 Insulation Board:

Insulation boards must be one of the following:

a. Sample Insulation Board, Sample II Insulation Board and Sample III Insulation Board, which are expanded polystyrene (EPS) complying with ASTM C578, Type I, and ASTM E2430. The boards are produced by a molder who participates in an approved third-party quality assurance program, and are labeled in accordance with Section 7.0 of this report. The boards include $^{1}/_{4}$-inch-deep-by-$1^{1}/_{2}$-inch-wide (3.2 mm by 38 mm) corrugations across their width.

b. Flat EPS insulation board from a molder with a current evaluation report showing conformance to ASTM E2430. The board must be labeled in accordance with the applicable report.

c. Flat EPS insulation board that complies with ASTM C578, Type I, and ASTM E2430. The board is produced by a molder who participates in an approved third-party quality assurance program, and is labeled in accordance with the applicable code.

3.3 Substrates:

- Gypsum sheathing complying with ASTM C1396 or ASTM C1177
- Fiber cement panels complying with ASTM C1186 and with the ICC-ES Acceptance Criteria for Fiber Cement Siding Used as Exterior Wall Siding (AC90)
- Fiber cement panels complying with ASTM C1325 and with the ICC-ES Acceptance Criteria for Reinforced Cementitious Sheets Used as Wall and Ceiling Sheathing and Floor Underlayment (AC376)
- Concrete-masonry complying with the code
- Concrete complying with the code
- Exposure 1 wood structural panels complying with DOC PS-1 or PS-2
- Brick masonry complying with the code

3.4 Water-resistive Barrier:

The barrier must be one of the following:

Page 1 of 4

FIGURE 2-3 Sample ICC evaluation report.

corrections must be made and the inspection approved before the wall finish material is applied to the studs. The inspections required by the IECC, therefore, do not add any additional inspection burden to projects.

Specific tests and inspections must be done to verify the work complies with the IECC. If the work does comply, the code official issues a notice of approval. This is usually a note or box on the inspection request to be marked "Approved" or "Rejected." The final inspection must be called for and approved before the building is occupied.

FIGURE 2-4 Approved plans stamp.

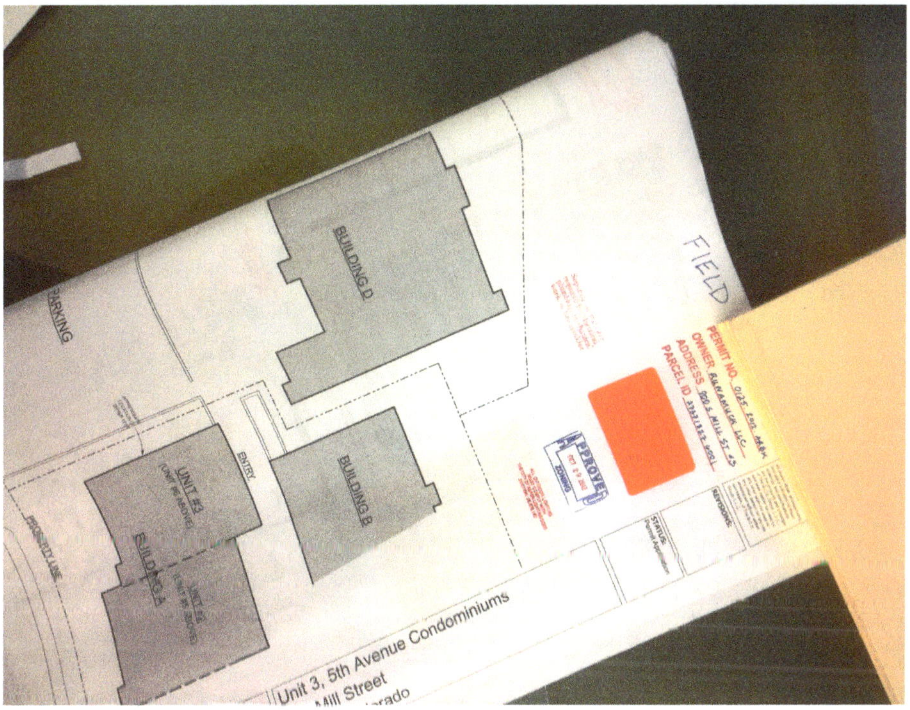

FIGURE 2-5 The field set of approved plans.

PART

II

General Commercial Energy Provisions

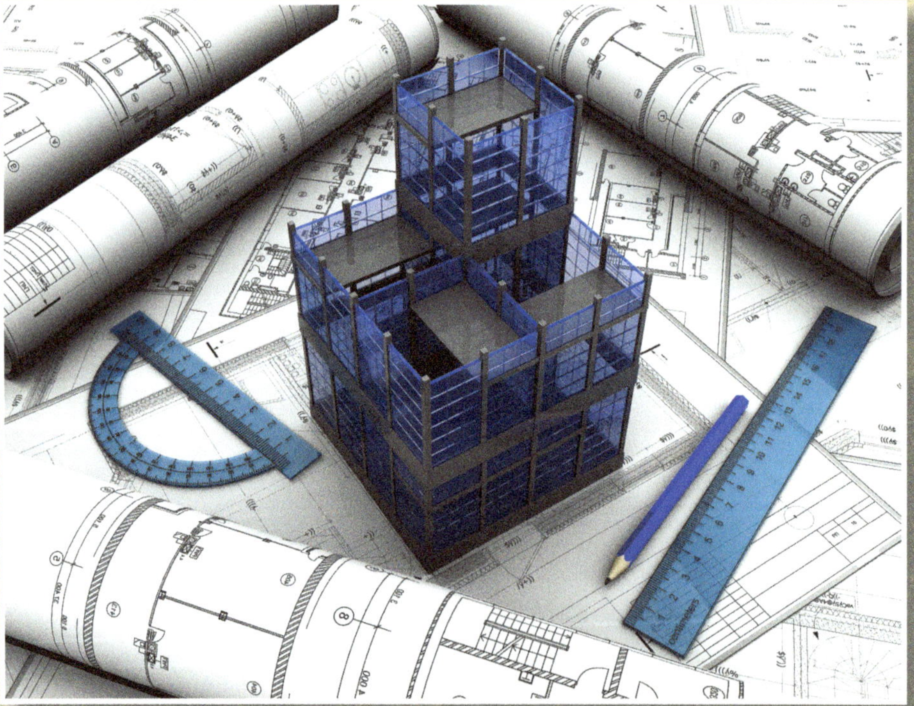

CHAPTER

3 General Commercial Energy Provisions

It is necessary to establish a framework to address the administration, application and enforcement of the commercial energy provisions set forth in the *International Energy Conservation Code®* (IECC®) and the *International Building Code®* (IBC®), among other codes. The code official's responsibilities regarding plan review and inspections, the architect's and engineer's responsibilities concerning document preparation and the contractor's responsibilities in regard to permits and inspections are clearly stated in the IECC. Sections C101 and C102 of the IECC, discussed in this section, govern the relationships and understandings between the building department authority and the design and construction community.

SCOPE

These code provisions apply to *commercial buildings* and building sites, systems and equipment as defined in the IECC. To fully understand the extent of the scope, it's important to know that the IECC, in some instances, utilizes specific unique definitions that are different from those found in the other International Codes® (I-Codes®), referenced standards and, in at least one instance, even between the commercial and residential provisions of the IECC. This becomes very apparent when discussing the scope of the code. The division between codes is set differently than the traditional IBC/IRC parameters. There are many buildings that will be constructed under IECC—Residential Provisions and the IBC for all other aspects.

IECC

C101.2 Scope. This code applies to *commercial buildings* and the buildings' sites and associated systems and equipment. **[Ref. C101.2]**
COMMERCIAL BUILDING. For this code, all buildings that are not included in the definition of "Residential building." **[Ref. C202]**
RESIDENTIAL BUILDING. For this code, includes detached one- and two-family dwellings and multiple single family dwellings (townhouses) and *Group* R-2, R-3 and R-4 buildings three stories or less in height above grade plane. **[Ref. C202]**

IRC

R101.2 Scope. The provisions of this code shall apply to the construction, *alteration*, movement, enlargement, replacement, repair, *equipment*, use and occupancy, location, removal and demolition of detached one- and two- family dwellings and *townhouses* not more than three stories above *grade plane* in height with a separate means of egress and their *accessory structures* not more than three stories above *grade plane* in height (see Figures 3-1 and 3-2). **[Ref. 2018 IRC R101.2]**

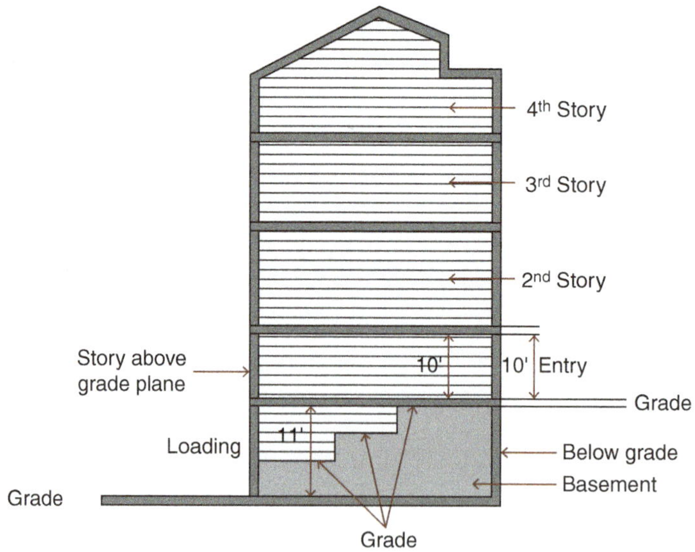

FIGURE 3-1 Story above grade plane.

FIGURE 3-2 A three-story apartment building is not regulated by the commercial energy provisions.

All of the examples above would be partially or completely regulated by both the IECC—Residential Provisions (GREEN) and the IBC for construction. Any R-1 or nonresidential occupancy would be within the scope of IECC—Commercial Provisions. (YELLOW)

EXAMPLE 1: Where a building includes both *residential building* and *commercial building* portions, each portion shall be separately considered and meet the applicable provisions of IECC—Commercial Provisions or IECC—Residential Provisions. The construction of the building would be completely within the scope of the IBC. **[Ref. C101.4.1]**

EXAMPLES 2 & 3: The IECC—Residential Provisions apply to all one- and two-family dwellings and townhomes, regardless of the number of stories above grade plane, while the IRC specifically limits its scope to three stories for one- and two-family dwellings and townhomes.

EXAMPLE 4: The IECC—Residential Provisions apply to an unlimited number of R-2, R-3 or R-4 units, as long as they are three stories or less above grade plane, while the IRC is limited to one- and two-family dwellings and townhomes three stories or less.

Determining the applicable IECC provisions is a process of elimination. If it doesn't meet the IECC definition of "Residential building," it's regulated by IECC—Commercial Provisions or ANSI/ASHRAE/IES Standard 90.1.

INTENT

The IECC utilizes a unique building classification system that is different from any other code.

In 2018 IECC Section C101.3, the statement of intent is clearly laid out in three sentences:

1. This code shall regulate the design and construction of buildings for the effective use and conservation of energy over the useful life of each building.

   This is the portion of this section that frequently gets the most attention. Some readers immediately fixate on the word "regulate" and stop reading at that point. Others will focus on the thought of "effective use and conservation of energy," which clearly should be included in the intent statement for an energy conservation code. And yet others will highlight the concept of "the useful life of each building." This sentence is like a three-legged stool; it needs appropriate attention to each phrase in order to achieve balance.

2. This code is intended to provide flexibility to permit the use of innovative approaches and techniques to achieve this objective.

   As important as the first sentence is in the Scoping section; the second, which emphasizes flexibility, is groundbreaking. In all of the I-Codes®, only the *International Energy Conservation Code*, *International Existing Building Code* and the *International Zoning Code* include the concept of flexibility in the Intent section. This code is intended to grant the design professional multiple choices, allowing the concepts of form and function to be pursued unhindered, while keeping cost in mind.

3. This code is not intended to abridge safety, health or environmental requirements contained in other applicable codes or ordinances.

   From a building safety point of view, this is the one single statement that keeps the IECC from becoming a deal breaker. There are rare occasions where strict observance of one or another IECC provision would cause a safety, health or environmental issue. The IECC is absolutely subservient to every applicable safety, health or environmental code, law and regulation, and any attempt to enforce a provision of the IECC in conflict clearly violates the intent of the code. **[Ref. C101.3]**

COMPLIANCE

Designers, builders and other permit applicants are solely afforded the opportunity to choose the compliance path, within the scope of the applicable codes. Code officials are limited to determining if the energy provisions applicable to a proposed project are met. Once again, the code official does not have the authority to demand one compliance method over another method that is acceptable within the framework of the code; the code official must accept any code-compliant path that is chosen by the permit applicant.

All plans, specifications and details—whether concerning new construction or the alteration or repair of an existing building—must comply with the minimum provisions of the code.

Some jurisdictions offer prescriptive worksheets that establish the requirements for building components and systems applicable in that location. This method is the easiest compliance path for simple commercial structures; requirements for each building element are listed on the worksheets and the plans are checked for compliance. The worksheets are specific to each climate zone and are often available from the state or local building department. **[Ref. C101.5.1]**

In addition, the U.S. Department of Energy (DOE) offers COM*check*™, a free, easy-to-use software program for demonstrating code compliance. The code official must approve the use of specific computer software such as COM*check*™. In such programs, the user inputs building areas, efficiencies and other specifications for the building envelope, mechanical systems and interior and exterior lighting systems. The software then generates a compliance report for the approved plans and a customized field inspection checklist. COM*check*™ compliance programs are available for editions of ANSI/ASHRAE/IES 90.1-2016, the IECC and specific state programs.

ALTERNATE METHODS AND MATERIALS

Innovation is encouraged in the IECC, therefore, materials or methods of construction that are not specified in the code but meet the intent of the applicable provision can be used if such are evaluated by the code official and approved before the building is permitted and inspected. The code official will consider the approval of the material or method based on the intent of the code provision that most closely describes the alternate. Section C102.1 was amended in the 2018 edition, with an additional sentence added: "Where the alternative material, design or method of construction is not *approved*, the code official shall respond in writing, stating the reasons why the alternative was not *approved*." This is a best practice that many code officials have observed throughout the years; however, it is now codified as mandatory, thus raising the bar. **[Ref. C102.1]**

Code Essentials

Alternative materials and methods and above-code programs:

- The permit applicant is responsible for supplying information supporting the proposed alternative.
- The code official is responsible for evaluating the information to determine equivalency and compliance.
- The code official may accept a program that exceeds the energy efficiency provisions of this code.
- If the code official rejects the alternative, the rejection must be in writing, stating the reason for the rejection. ●

EXISTING BUILDINGS

The 2018 IBC defines "Existing building" as follows: "A building erected prior to the date of adoption of the appropriate code, or one for which a legal building *permit* has been issued."

The IECC does not require untouched portions of a legally constructed (permitted) existing building or building system to be upgraded in order comply with the energy code provisions for new buildings. Buildings constructed without a required permit are not covered by this exemption. Any proposed work in an existing building that requires a permit must meet the IECC provisions.

2018 IECC Section C502.1 offers an alternative method for compliance when placing an addition on an existing building. "An *addition* shall be deemed to comply with this code if the *addition* alone complies or if the existing building and *addition* comply with this code as a single building." It should be noted that this could possibly allow someone to build a below-code *addition* on an existing high-performance building, as long as the existing building and addition comply as a single building. **[Ref. C501 & C502]**

HISTORIC BUILDING

The definition of "Historic building" in Section 202 lists the criteria for determining if a building meets that status.

HISTORIC BUILDING. Any building or structure that is one or more of the following:

1. Listed, or certified as eligible for listing by the State Historic Preservation Officer or the Keeper of the National Register of Historic Places, in the National Register of Historic Places.
2. Designated as historic under an applicable state or local law.
3. Certified as a contributing resource within a National Register-listed, state-designated or locally designated historic district.

Any building that meets these criteria may be allowed the relaxed IECC provisions for historic buildings.

WORK REQUIRING PERMITS

Most existing buildings are not considered historic, so any new work in these buildings—such as additions and renovations—must comply with the energy code provisions required for new construction. Parts of the building not affected by the new construction do not need to meet code provisions. This approach, "if you work on it, then it must comply," is consistent in all the I-Codes.

Although new work must comply with the code, there are specific exceptions—for example, projects or renovations that will not increase the building's energy use are exempt. However, installing a new window assembly in an existing window opening requires the window to meet the same requirements for solar heat gain coefficient (SHGC) and *U*-factor ratings as would be required for new construction. Typical energy-upgrade projects that include window replacement require permit application, plan review, permit issuance and inspection.

Commercial tenant spaces that rearrange lighting would typically require an electrical permit in many jurisdictions; however, if more than half of the lighting fixtures in the space are replaced or the new lighting design increases energy use in the space, then compliance with the IECC—Commercial Provisions would be mandatory. Compliance is usually confirmed with a lighting plan and fixture schedule. **[Ref. C503.1]**

CHANGE IN USE OR OCCUPANCY

Commercial tenant occupancy changes may trigger review of energy code compliance. Although residential apartment tenants come and go, the space is still an apartment. As commercial tenants come and go, the code requirements may change due to the new tenant's finish work for the building space. For example, the tenant finish work does not require energy code compliance if no alterations are made to the lighting

You Should Know

When calculating the "space" to determine if IECC compliance is required for lighting, the definition of "Enclosed space" in Section 202 provides guidance: "**ENCLOSED SPACE.** A volume surrounded by solid surfaces such as walls, floors, roofs, and openable devices such as doors and operable windows." **[Ref. C202.2]**

With this definition in mind, a large area with a newly created small office built in one corner provides a good example. Because less than half of the larger space has been modified, no IECC provisions are applicable. The new small office space has separate lighting with individual controls for 100 percent of its space and the lighting and lighting controls in this space (only) will be regulated by the IECC. ●

systems, but if changes are made, the new lighting system cannot use more power than is allowed for new construction. This may happen when a retail tenant moves into an existing office space and installs merchandise lighting. The new lighting displays may use more power than office lighting, but cannot use more than is allowed by the code for new construction. An example of a total and complete change of use and occupancy is a house converted into an office space or business. **[Ref. C505]**

Similar to the requirement for a change in occupancy is when any previously unconditioned space becomes heated or cooled. A warehouse or storage area that becomes an office or work area must comply with all the provisions of the code, which may require insulation changes as well as the replacement of doors and windows. **[Ref. C503.2]**

All commercial building uses are regulated by the provisions in this part of the code. In buildings not more than three stories tall that contain both commercial and residential spaces, the office and retail uses must comply with these commercial provisions. Residential spaces in a building not more than three stories high are regulated by the residential provisions of the code. Hotel or apartment buildings four stories or more, even though the entire use is residential, must meet the commercial energy provisions. **[Ref. C101.5]**

CHAPTER 4

Administration and Enforcement

PREPARING THE PLANS

One responsibility of the permit applicant is to coordinate the skills of the design team to prepare a complete set of construction documents for the project, which will then be reviewed by a plans examiner for compliance with the *International Energy Conservation Code*® (IECC®).

It should be obvious that the design professionals and contractors who understand the specific provisions and requirements of the code will produce better products with many benefits.

- One of the largest benefits is that the IECC is intended to provide flexibility for design and construction. This provides options in form, function and budget that are frequently overlooked by the less informed.
- Another benefit is that better quality products are approved more quickly.

In some cases, a simple set of building plans with details and schedules will describe a project sufficiently to demonstrate compliance with the code. However, for many commercial building projects, a complex set of documents—including mechanical, electrical and plumbing plans—must be prepared by registered design professionals. In the early phase of design development, it is important to know if the jurisdiction of the proposed building requires the plans to be prepared by a registered design professional. Many jurisdictions have incorporated a threshold level for signature and seal into their laws. The code official is also authorized to waive this requirement if it is determined that a registered design professional's stamp and seal are not required to confirm compliance with the code. **[Ref. C103.1]**

Section C103.2 provides a detailed list of the information that is required to be provided on the construction documents. Elements of the building envelope, mechanical system drawings and schedules, service water heating information, and electrical and lighting system drawings and luminaire (lighting) schedules must be included to make a complete set of drawings for review of energy code compliance.

Once a compliance path is chosen, the required details to demonstrate compliance will be specific to the chosen (prescriptive or performance) compliance path. **[Ref. C103.2]**

Code Essentials

Air leakage is the largest single cause of energy loss in many buildings. An effective air barrier is the best way to control air leakage. ●

THE CONSTRUCTION DRAWINGS AND DOCUMENTS

The thermal envelope and air sealing

The construction documents must clearly indicate the nature and extent of the proposed work. The details include insulation materials, window and door schedules, mechanical system design, lighting fixture schedule, location of daylight zones and air-sealing details. **[Ref. C103.2]**

The building plans must include building elevations depicting window, door and skylight areas and wall sections showing the type of insulation, its thermal performance and its thickness. This applies to a single-story retail building, a four-story apartment building and a 40-story office building. The thermal envelope must be continuous on every element of the exterior wall, roof and fenestration detail. The design documents must have sufficient detail to confirm that all six sides of the building cube address insulation continuity. The thermal envelope must be completely described by the submittal documents. This description may be on the drawings, in notes or specifications, on tables, or in software printouts. Please note that the submittals must be consistent when referring to a specific element, i.e., the notes calling for R-40 batt fiberglass insulation in a 2-inch cavity shown on the drawings. **[Ref. C103.2.1]**

These wall and roof sections will list the thermal performance—typically in *R*-values or *U*-factors of the insulation, and the window and door schedules will list the *U*-factors and solar heat gain coefficients (SHGCs) of the fenestration specified for the project. Wall components and framing must be identified separately. This is important, as the insulation and material requirements for wood- versus metal-framed walls are very different. Building plan elevations indicating window and door sizes and locations, and a roof plan showing skylights, must agree with the schedules, notes and callouts. Caulking and sealing details for window and door frame installation must be included in the construction documents submitted for plan review. These notes and details are typical of how all exterior joints are to be sealed.

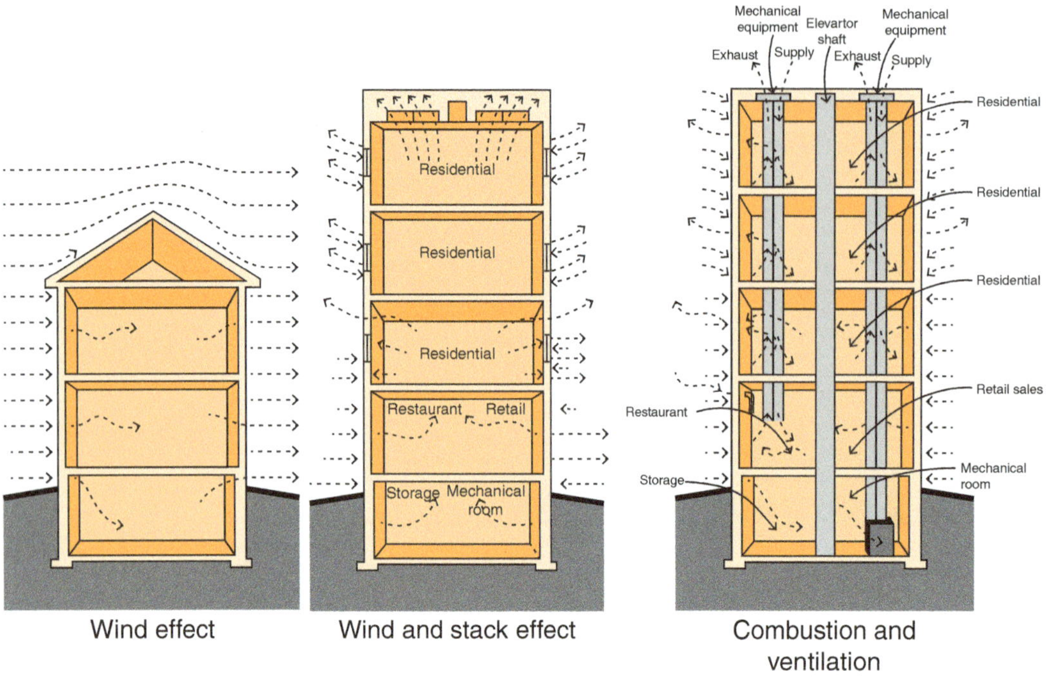

FIGURE 4-1 Examples of air infiltration.

You Should Know

Commercial building exterior wall assemblies often present very different challenges than those typical in residential construction. For example, the structural design requirements in a high-rise building include a strong *and* flexible frame design. The exterior cladding system may be made from glass, metal, stone or a variety of other materials to create an architecturally pleasing face for the structure. The thermal barrier fits into wall sections between the interior wall finish and the exterior cladding.

Material choices for insulation must comply with the *R*-values or *U*-factors required by the energy code, and the flame-spread and smoke-development characteristics must comply with the applicable provisions in the *International Building Code* (IBC). Section 720 of the *International Building Code*® (IBC®) contains the requirements for thermal insulation in walls, roofs and around pipes. IBC Section 1406 contains the provisions for Metal Composite Material (MCM), which was used on the now-infamous Grenfell Tower, where 72 people were killed as the MCM cladding burned, while Section 1407 specifically addresses exterior insulation and finish system (EIFS) materials, construction and inspection. The requirements for insulation installed above the roof deck are noted in IBC Section 1508. Specific provisions for foam plastic material and installation are found in IBC Section 2603.

Fire-resistance-rated construction is an essential element of commercial building design and fire protection. The flame-spread restriction for materials helps control the growth of fire. Building occupants and rescue personnel must deal with smoke in a fire situation; establishing maximum limits for the smoke-development properties of insulation materials maximizes time for occupants to exit the building and for fire fighters to react to the situation. The fire-resistive properties of building materials and insulation must meet the requirements for the life-safety provisions in the IBC. The intent is that the registered design professional team will select insulation material and provide building design details that comply with both the IECC and the IBC. ●

Air infiltration is the unintentional and uncontrolled movement of outside air into the building. The air usually moves through cracks and holes in the building envelope and doors used for egress. These same air-movement pathways allow for *air exfiltration*—inside air moving through the building envelope to the outside. The three primary aggravating conditions on air leakage are wind, the stack effect and mechanical equipment in the building (Figure 4-1). Wind creates pressure differences on opposite faces of the building. The stack effect creates unequal pressures resulting from the differences in density between heated interior air and exterior cold air or cooled interior air and warm exterior air. This is magnified by the height of the building. HVAC exhaust fans and systems and supply fans and systems can also lead to pressure differences in a building. A tight building envelope helps control air movement and will help lower heating and cooling costs, make the building more healthy and comfortable to live and work in and improve the performance of the HVAC system.

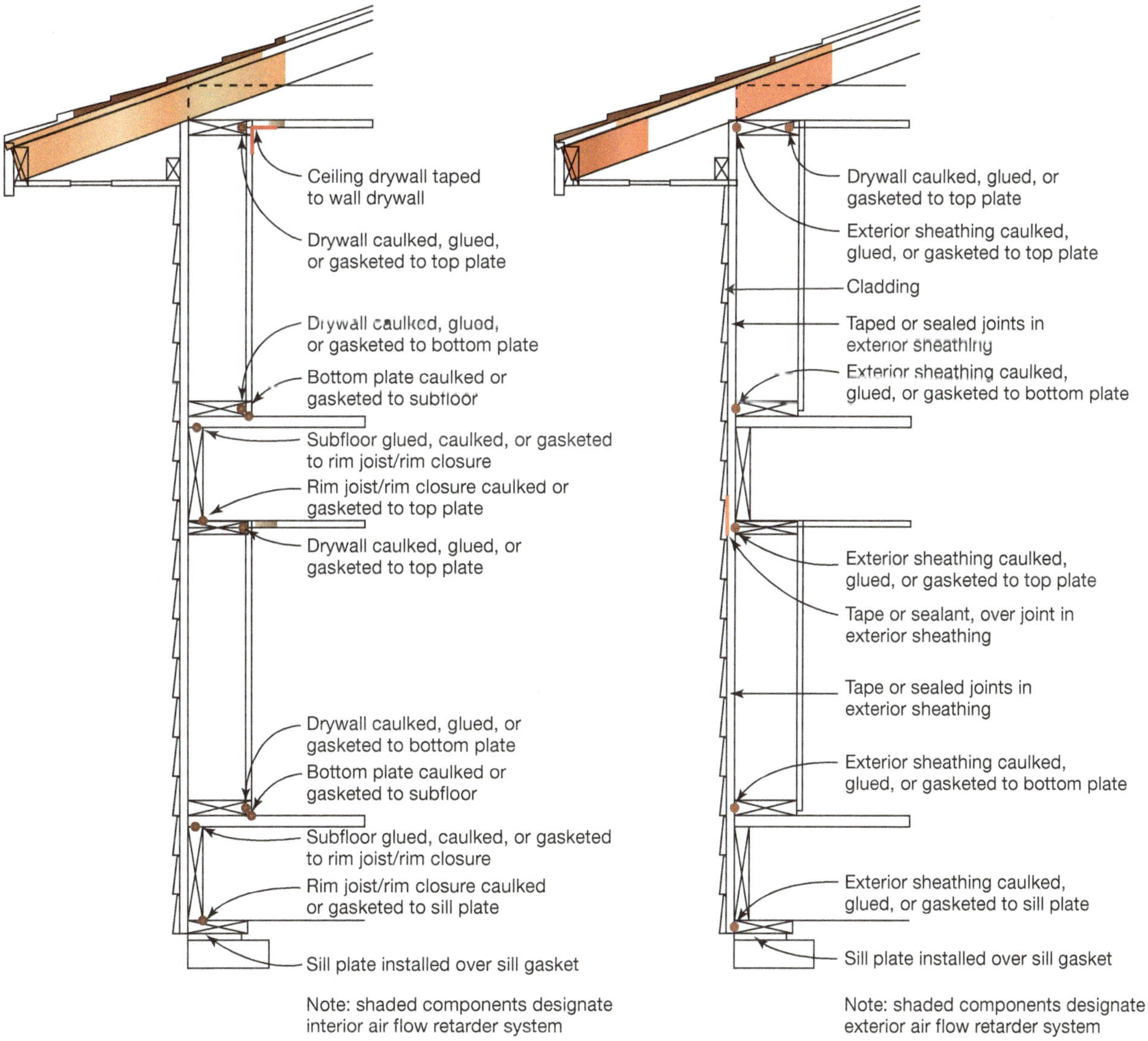

FIGURE 4-2 Envelope air sealing. *(Courtesy of Building Science Corporation).*

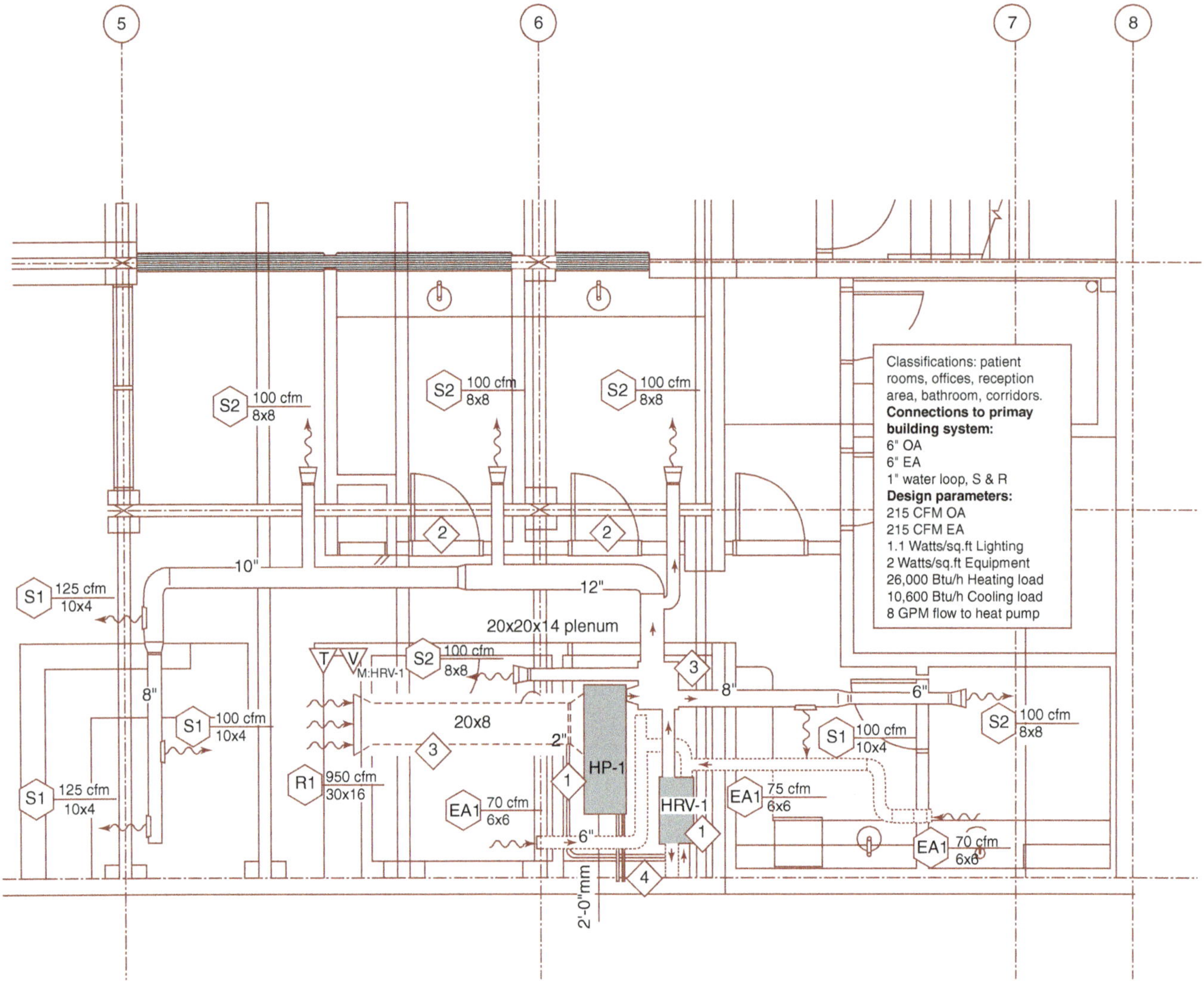

FIGURE 4-3 Mechanical plan.

Air leakage controls must be clearly addressed in the building plan documents. Again, the typical commercial building exterior wall is a complex assembly. All exterior joints, cracks and holes are to be gasketed, caulked, weather-stripped, or sealed to create a continuous air-barrier layer. The air-sealing materials must be called out on the plans and be compatible with the building materials in the joints and holes where they are applied. Details, notes and sections in the building plans must address the continuity of the air sealing for the entire building envelope (Figure 4-2).

Mechanical system

Mechanical system design is particularly important in commercial construction. Most mechanical systems for commercial uses are sophisticated, complex and designed by mechanical engineers (Figure 4-3). The documents require a statement describing design criteria and mechanical equipment schedules, including the types and efficiencies of the HVAC equipment and the systems controls.

The building ventilation system details must include fan motor horsepower and control system operation (Table 4-1). Duct sealing is important to minimize uncontrolled conditioned air leakage in plenums and interior building spaces. Duct-sealing materials must be specified to demonstrate their compatibility with the duct materials. Air ducts and pipes carrying conditioned air or water are required to be insulated. The insulation minimizes heat loss or heat gain as the HVAC system delivers the conditioned air or heated or chilled water throughout the building.

TABLE 4-1 Mechanical equipment schedule

Mechanical Schedule

Id	Description	Location	Manufacturer & Model	cfm	esp	Volt-age	Phase	HP/ watts	FLA
EA-1	Exhaust air grille, with integrated damper	Various	60 DAL-L , coordinate finish with Arch	–	–	–	–	–	–
HP-1	Horizontal water-to-air heat pump, condensate drain required	High in bathroom space	Water Furnace ND-H-026, with thermostat, 2" Merv 11 filter, hang with vibration isolation springs.	950	~ 0.2	230	1	373 watts	14.2
HRV-1	Heat recovery ventilator, condensate drain required	High in bathroom space	SHR 2004, EDF2 controller, hang with vibration isolation springs.	215	~ 0.3	115	1	150 watts	2.1
R-1	Return air grille	Various	60-F, coordinate finish with Arch	–	–	–	–	–	–
S-1	Spiral duct supply grille, with air scoop/damper	Various	SDGE-AS, air scoop, size for duct diameter (see plans), coordinate finish with Arch	–	–	–	–	–	–
S-2	Supply air grille, with integrated damper	Various	22DAL-ED-L, exposed duct connection, coordinate finish with Arch	–	–	–	–	–	–

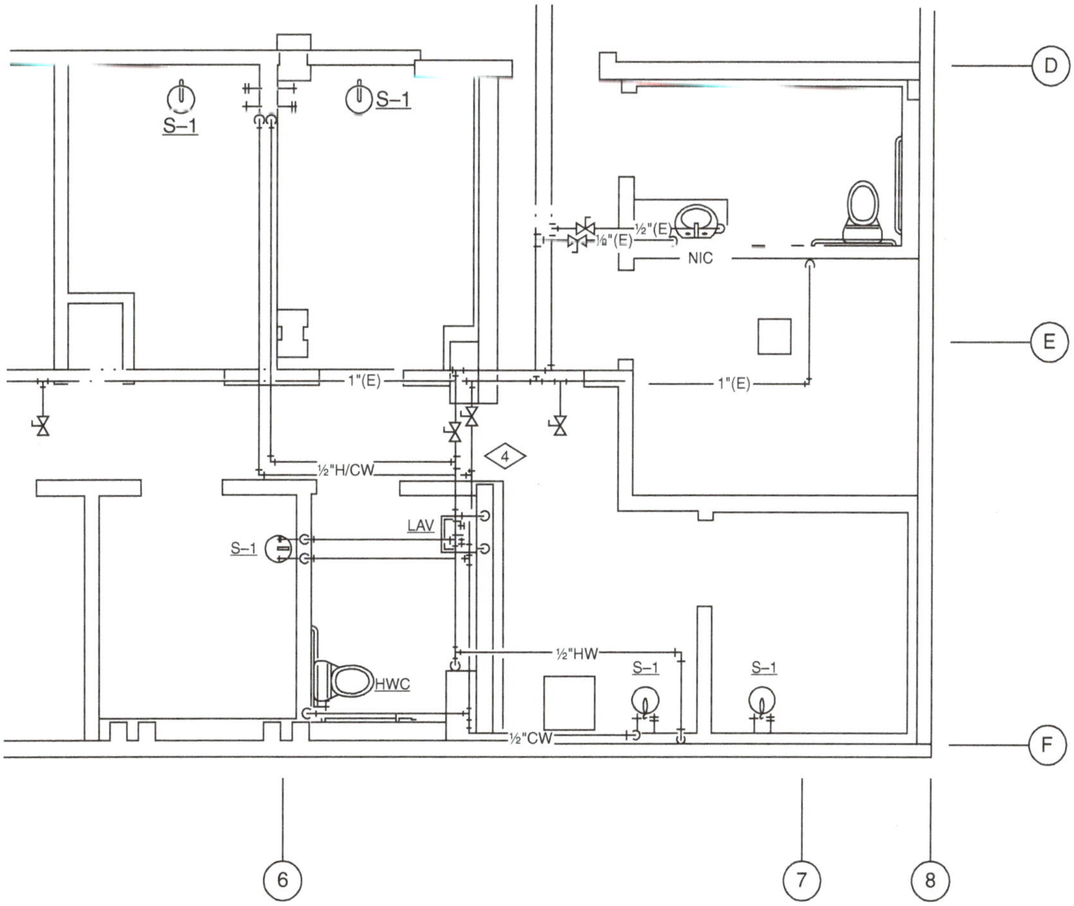

FIGURE 4-4 Plumbing plan.

TABLE 4-2 Plumbing fixture schedule

Plumbing Fixture Schedule			
Key	**Description**	**Fittings/accessories**	**Manufacturer/catalog #**
Lav	Square Wall-mounted Lavatory, ADA Compliant, Center Hole, White	Deck Mount Single Hole, Pop-up Drain, Offset Tailpiece, Insulated Drain Piping	Porcher 26020 - Center Hole Lavatory Delta 22c801
Hwc	Handicap Water Closet, Dual Flush Tank Type, Floor Mounted, Siphon Jet, Elongated Bowl, 1.0/1.65 Gallon, 16.5″ Rim Height	Extra Heavy Duty Open Front, Solid Plastic Seat, Check Hinge	American Standard 2888.216 (Flo-wise) 5322.011 (Seat With Cover)
S-1	Single Bowl Undermount Sink, 20 Guage Stainless	Deck Mount Single Hole, Pop-up Drain	Kindred Qsu1816/8n Delta 22c801

Plumbing Fixture Pipe Sizes		
Key	Hw Size	Cw Size
H/wc (flush tank)	-	1/2″
Lav	1/2″	1/2″
S-1	1/2″	1/2″

All pipe sizes as indicated except where noted.

Service water heating

Service water heating systems require separate design details and schedules for equipment efficiencies and control systems. The system piping layout, equipment location and control operation information must be included in the design documents (Figure 4-4). Water heating equipment size and efficiency schedules are required (Table 4-2). Calculations may be included to demonstrate compliance with specific code provisions such as the maximum allowable pipe volume. Supplying the information that the plan reviewer needs to determine compliance will speed the process of plan approval and permit issuance.

Lighting and controls

Interior and exterior lighting systems and controls must be included in the design documents and must demonstrate compliance with the provisions of the IECC. The interior lighting system and control strategy is usually described in the reflected ceiling plan (Figure 4-5). A lighting fixture schedule, including the wattage of each luminaire, is required (Table 4-3).

PLAN REVIEW

The code official is charged with reviewing the construction documents for code compliance. [Ref. C103.3] It is not just the building plans but the entire suite of documents—including schedules,

Code Essentials

A strategy for moisture and vapor control must be addressed in the building plan details. The requirements in IBC Section 1404.3 must be incorporated in the building plan wall section details. Exterior walls must be covered with materials approved as weather coverings and, in certain climate zones, a vapor retarder. It is important to be familiar with and incorporate these provisions in the plan set. ●

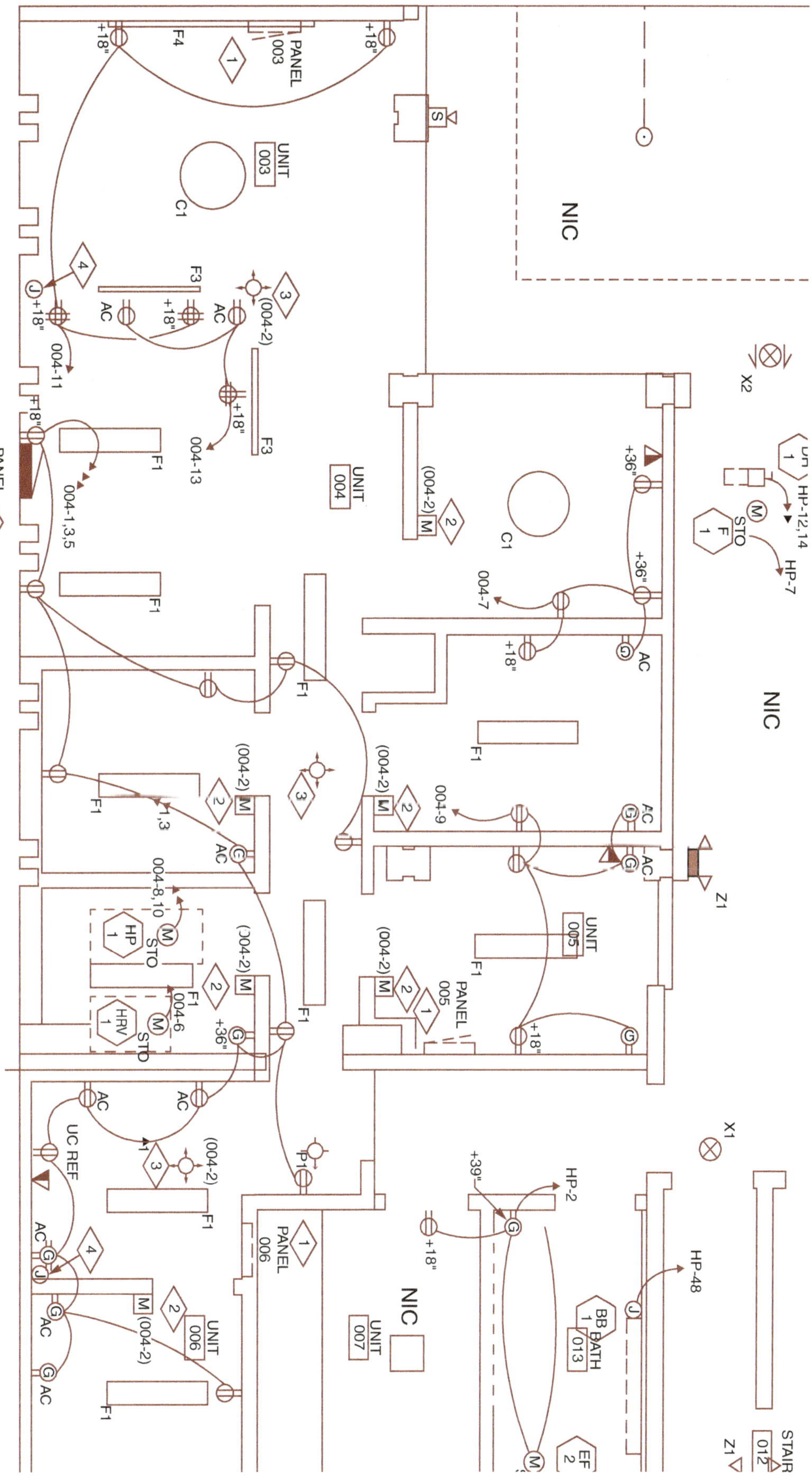

FIGURE 4-5 Electrical plan.

TABLE 4-3 Luminaire schedule

Luminaire schedule

Key	Lamp	Description	Ceiling (depth)	Manufacturer/#	Volt
C1	42w Gu-24 Cfl	Round contemporary chandelier, coordinate finish with architect	Pendant, verify height with architect	Lampa Satellite Chandelier	120
F1	(2)F32/t8 Spx35	Linear fluorescent pendant	Pendant, verify height with architect	Metalumen Planar S2e	120
F3	Led	4' Straight & narrow led, direct, white, powder coated acrylic	Pendant, verify height with architect	Cooper Lighting Neo-ray 22-d-p	120
F4	Led	8' Straight & narrow led, direct, white, powder coated acrylic	Pendant, verify height with architect	Cooper Lighting Neo-ray 22-d-p	120
P1	75w Par 30	Line voltage direction spotlight, white	Surface	Artemide Architectural Starship Surface	120

Section 2: Interior lighting and power calculation

A Area category	B Floor area (Ft^2)	C Allowed watts/ft^2	D Allowed watts (B X C)
Healthcare clinic	1110	0.90	999
		Total allowed watts =	999
Area category exemption qualifications			

Activity area	Total wattage Pre-alt.	Total wattage Post-alt.	Total Pre-alt. Fixtures	# Fixtures Repl./added

Section 3: Interior lighting fixture schedule

A Fixture Id : Description/lamp/wattage per lamp/ballast	B Lamps/ fixtures	C #of fixtures	D Fixture watt.	E (C X D)
Healthcare clinic (971 sq.ft.)				
Linear fluorescent 1: F1: suspended flourescent /48" T8 32w/electronic	2	10	64	640
Compact fluorescent 1: C1: satellite chandelier/triple 4-pin 42w/electronic	1	2	42	84
Led 3: P1: ceiling mounted monopoint/other	1	1	23	23
Led 1: F3: suspended led/other	1	2	45	90
Led 2: F4: wall mounted led/other	2	1	90	90
			Total proposed watts =	927

calculations and specific product information—that is subject to review for code compliance. The final act in permit issuance is to stamp the approved document(s) "Reviewed for Code Compliance." One set of approved plans is kept in the building department office and an identical set is kept at the work site. This approved set of construction documents cannot be altered or changed without the code official's approval. **[Ref. C103.3.1]**

Officials will visit the work site for field inspections, and the construction must agree with the approved plans. Any changes to the approved plans must be resubmitted to the code official for further review, approved and then the re-approved plans must be returned to the field before inspections are called for that portion of the work. **[Ref. C103.4]**

INSPECTIONS

Building construction work that requires a permit also requires inspection. Specific inspections are listed in Section C105 of the IECC and complement those listed in the other I-Codes. The periodic inspections required by the normal construction process already incorporate many provisions of the energy code in the inspection schedule, so the list of required inspections that have been included in the 2018 IECC should not add a substantial burden to the building owners, contractors or inspectors. For example, the foundation inspection requires perimeter and below-grade wall insulation to be in place before backfilling. During the framing or by special inspection for commercial construction, field staff will verify fenestration compliance and air sealing. The insulation inspection verifies that materials of the required *R*-values are in place and properly installed.

Mechanical, plumbing and electrical inspections are required to verify compliance with the approved plans. The installations must be done safely and must comply with the requirements of the applicable codes as well as the approved plans. The schedules indicate the HVAC efficiency requirements; system controls and duct size, location and insulation. Service water heating inspections include verifying that equipment efficiencies match those documented on the plans, temperature controls are installed as approved and pipe insulation is installed as required. Electrical and lighting inspections verify the number and type of fixtures installed according to the approved plans. Interior and exterior controls, switches and timers must be installed according to the approved plans. Specific energy code inspections include those listed in Table 4-4. **[Ref. C105]**

Contractors and tradespeople are responsible for knowing the local rules regarding field inspection expectations, as jurisdictions may have their own standards and amendments.

FEES

Every jurisdiction that adopts building codes establishes a fee schedule. Fee amounts vary, so it is important to get specific information from the jurisdiction in which the proposed project is located. Any and all applicable fees must be paid before a specific permit is issued. Energy code compliance may require building, mechanical, plumbing and electrical permits. Any work that needs a permit should not begin before the permit is issued. **[Ref. C104.1]** Work without a permit is subject to additional fees added to the permit fee. **[Ref. C104.3]** It is not a good idea to start work before a permit is issued, as the work may not comply with the applicable code provisions. This may be a costly mistake, resulting in more time, money and materials than necessary.

TABLE 4-4 Required inspections

Section	Inspection	Details
C105.2.1	Footing and foundation inspection	Inspections associated with footings and foundations shall verify compliance with the code as to *R*-value, location, thickness, depth of burial and protection of insulation as required by the code and approved plans and specifications.
C105.2.2	Framing and foundation inspection	Inspections at framing and rough-in shall be made before application of interior finish and shall verify compliance with the code as to types of insulation and corresponding *R*-values and their correct location and proper installation; fenestration properties (*U*-factor, SHGC and VT) and proper installation; and air leakage controls as required by the code and approved plans and specifications.
C105.2.3	Plumbing system	Inspections at plumbing rough-in shall verify compliance as required by the code and approved plans and specifications as to types of insulation and corresponding *R*-values and protection; required controls; and required heat traps.
C105.2.4	Mechanical system	Inspections at mechanical rough-in shall verify compliance as required by the code and approved plans and specifications as to installed HVAC equipment type and size; required controls, system insulation and corresponding *R*-value; system and damper air leakage; and required energy recovery and economizers.
C105.2.5	Electrical system	Inspections at electrical rough-in shall verify compliance as required by the code and approved plans and specifications as to installed lighting systems, components and controls; and installation of an electric meter for each dwelling unit.
C105.2.6	Final inspection	The building shall have a final inspection and shall not be occupied until approved. The final inspection shall include verification of the installation and proper operation of all required building controls, and documentation verifying activities associated with required building commissioning have been conducted and findings of noncompliance corrected. Buildings, or portions thereof, shall not be considered for a final inspection until the code official has received a letter of transmittal from the building owner acknowledging that the building owner has received the Preliminary Commissioning Report as required in Section C408.

ENFORCEMENT

A stop work order is the ultimate enforcement tool (Figure 4-6). The code official has the authority to stop work on any part of a project if the work regulated by the code is dangerous, unsafe or not code compliant. **[Ref. C108.1]** The stop work order must be in writing, is issued to the owner or person doing the work and must be specific to the noncompliant work ordered to be stopped. **[Ref. C108.2]** If the stop work order is ignored, the code official may prosecute the person responsible—an action that generally involves a citation in a municipal court. The amount of the fine is determined by each jurisdiction and written into the code at its adoption. **[Ref. C108.4]**

LEGAL NOTICE

WHEREAS, VIOLATIONS OF: Village of Oak Brook Code
have been found on these premises.

☐Section 105.1 Permit Required. ☐Section 113.1 Unlawful Acts

☐Section 114.3 Unlawful Continuance ☐Other__________________

It is hereby ordered in accordance with the above Code that all persons cease, desist from and

STOP WORK

at once pertaining to ANY construction, alterations or repairs on these premises, known as:

____

UNTIL ALL REQUIRED PERMITS HAVE BEEN OBTAINED AND PROPERLY DISPLAYED.
Persons acting contrary to this order, or removing/mutilating this notice are liable to arrest unless the action is authorized by the Village of Oak Brook .

______________________ ______________________
Inspector Date
Phone 630/368-5101 www.oak-brook.org Fax 630/368-5128
20100323

FIGURE 4-6 Typical red tag form.

BOARD OF APPEALS

The board of appeals exists to allow a means to appeal decisions made by the building official. Members of boards of appeal are usually civic-minded architects, engineers, insulators, lighting designers or mechanical, plumbing or electrical contractors. The board of appeals decides whether the appellant or code official's action is consistent with the intent of the code. For example, it may be the case that the IECC does not apply to a specific project condition, or that an equally good or better alternative approach to a proposed construction method is rejected by the code official. It is the broad understanding of the energy code that qualifies board members to hear such appeals. **[Ref. C109.3]**

Note that it is the *intent* of a specific code provision that generates the basis for an appeal. An appeal may be heard when an appellant—perhaps an owner, designer or tradesperson—does not agree with the code official's decision or order. The appeal may concern a design or field inspection decision that the appellant believes is incorrect based on the true intent of the code provision or rule as applied by the code official. **[Ref. C109.1]** The board hears the appeal, makes a decision and records the findings in writing. In no case may the board waive any requirements of the code. **[Ref. C109.2]**

DEFINED TERMS

Many code experts agree that the defined terms section is the most important part of any code or standard. The 134 specific words and terms listed in Chapter 2 of the IECC—Commercial Provisions define what code users agree to understand as the meaning of such terms when written and appearing in the code provisions. **[Ref. C201]** For example, the terms *access to, ready access to, building, listed, repair, storefront* and *zone* all are defined. Any defined word or term appears in *italics* in the code text. In reading, interpreting and applying code provisions, it is essential to understand that some of the definitions in the IECC are unique to the extent that some differ even between the commercial and residential provisions. **[Ref. C202]** This is the basis for civil discussion and deeper understanding of the energy code provisions.

Code Essentials

The user must first go to the appropriate (Commercial or Residential) IECC Chapter 2. Terms that are not defined in the IECC, but are defined in the *International Building Code, International Fire Code, International Fuel Gas Code, International Mechanical Code, International Plumbing Code* or the *International Residential Code,* shall have the meanings ascribed to them in those codes. All other terms shall have ordinarily accepted meanings such as the context implies. **[Ref C201]** •

PART III

Specific Requirements For Commercial Buildings

CHAPTER

5 Requirements for Commercial Buildings

Design criteria must be established to guide users to regional and specific climate and moisture data most appropriate to the area where the building will be constructed. The climate zone map references:

- every county in the United States,
- the territories of Guam, Puerto Rico, American Samoa,
- the commonwealth of Northern Mariana Islands and
- the Virgin Islands.

In addition, design criteria for any location in the world can be determined from the international climate zone definitions. **[Ref. Table C301.3(1)]**

CLIMATE ZONES

The climate zone map is a simple, easy-to-understand graphic depiction of historical weather data (Figure 5-1). The colorful visual map is supplemented by Table C301.1 in the *International Energy Conservation Code*® (IECC®), which lists climate zones, moisture regimes and warm-humid designations by state, county and territory. This section also defines climate types (moisture regimes) and thermal criteria so that any place in the world with reliable weather data can be assigned a climate zone and climate type. **[Ref. C301.1]** This information is the starting point in energy-efficient building design. All of the prescriptive building envelope provisions are based on weather data for the building's proposed location. The climate zone map is recognized in the development community as the important base document to fulfill the intent of the code to "regulate the design and construction of buildings for the use and conservation of energy over the life of each building."

Heating Degree Days (HDD) and Cooling Degree Days (CDD) climate data define the separation criteria of the zones. Both are reliable indicators for predictable energy use and determine insulation and fenestration requirements and the performance of heating and cooling equipment. A basic knowledge of these terms is helpful in understanding the need for the prescriptive provisions of the code. The greater the HDD, the better the building must be designed to perform in cold weather. The greater the CDD, the better the building must be designed to perform in hot weather. Insulation *R*-values are relaxed in warmer climates, while values for solar heat gain coefficient (SHGC) are relaxed in colder climates. Building envelope components are adjusted to fit the conditions of the building's location.

You Should Know

The climate zone maps and tables in the 2018 IECC and ANSI/ASHRAE/IESNA 90.1 are not interchangeable.

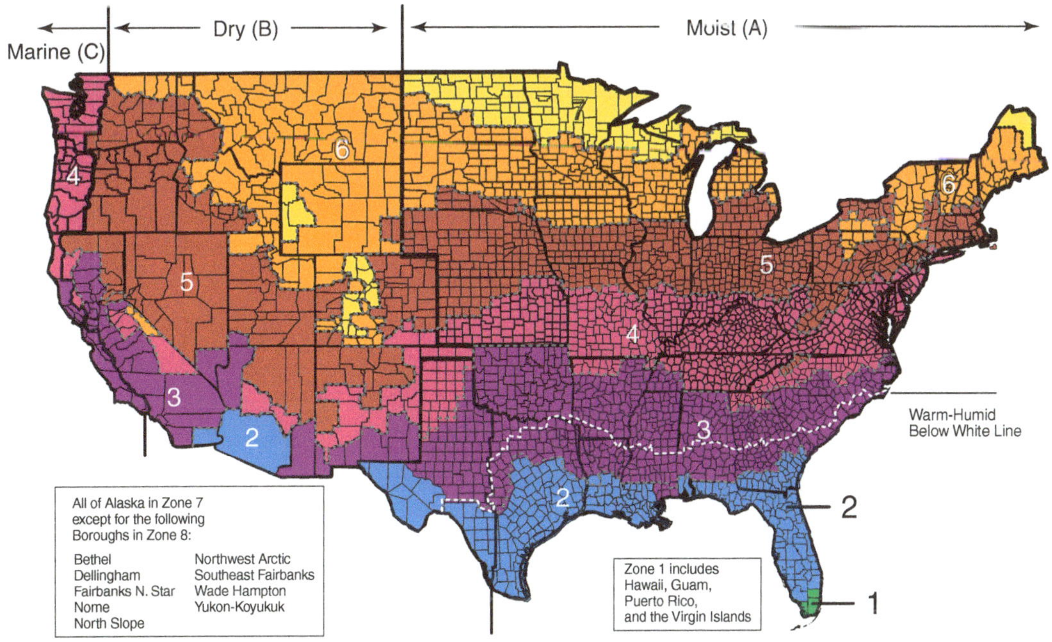

FIGURE 5-1 Climate zone map.

FIGURE 5-2 Marine climate conditions.

FIGURE 5-3 Moist climate conditions.

FIGURE 5-4 Dry climate conditions.

The reader needs to be aware that Section C401.2 allows the user to demonstrate compliance by meeting the requirements of ANSI/ASHRAE/IESNA 90.1-2016. However, the climate map for that standard has changed significantly—relocating approximately ten percent (10%) of the counties in the United States into a new climate zone, thus creating the very real possibility that a project will be in one climate zone when meeting the provisions of the IECC and another when meeting those of Standard 90.1.

FIGURE 5-5 Tropical climate zone.

The climate zone map also recognizes the moisture-regime designations of marine, dry and moist conditions, as well as warm-humid counties (Figures 5-2, 5-3 and 5-4). The designations are based on the combination and duration of seasonal temperatures and precipitation. Provisions for energy recovery ventilation and economizer mechanical equipment are determined by the climate and moisture-regime designations. The tropical climate zone includes the islands between the Tropic of Cancer and the Tropic of Capricorn. **[Ref. 301.4]**

DESIGN CONDITIONS

Interior design temperatures are established in the code to properly size heating and cooling mechanical equipment. The maximum design temperature for heating equipment is 72°F (22°C). The minimum design temperature for cooling is 75°F (24°C). **[Ref. C302.1]** It is very important to note that these criteria provide essential input for mechanical equipment design; however, they do not establish the interior comfort or livability provisions addressed in other building codes.

Code Essentials

Section 1203.1 of the *International Building Code* (IBC) and Section R303.10 of the *International Residential Code* (IRC) establish the minimum interior comfort temperature for heating. ●

INSULATION MATERIALS

Insulation materials used in the building thermal envelope must somehow be marked and identified to confirm compliance with the approved plans. **[Ref. C303.1]** Each piece of batt insulation 12 inches or greater in width must be marked by the manufacturer (Figure 5-6).

FIGURE 5-6 Unfaced insulation marked on bag.

If the product is not stamped at the factory, the installer must provide a certificate listing the insulation values and information for each area in which the unmarked insulation is installed in the building thermal envelope. Blown or sprayed fiberglass and cellulose wall insulation require certification from the installer specifying thickness and *R*-value details. Insulated siding shall have the *R*-value labeled on the product package (Figure 5-7). The certificate for sprayed polyurethane foam (SPF) requires the *R*-value of the installed thickness of the foam (Figure 5-8). In all cases, the installer must sign, date and post the certification at the job site. **[Ref. C303.1.1]**

You Should Know

For fiberglass or cellulose blown-in or sprayed insulation, the certification must include six characteristics, one of which is installed density. ●

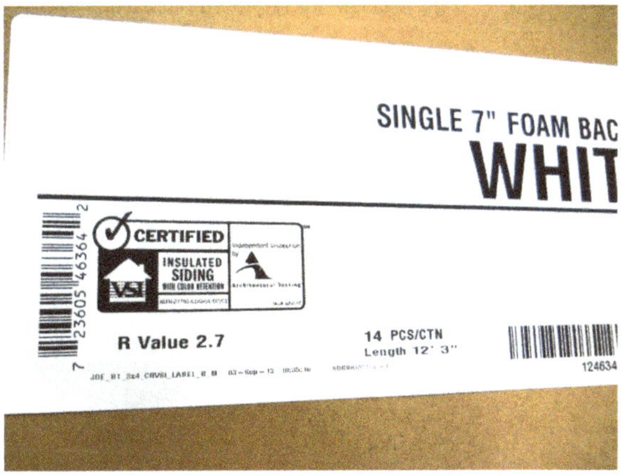

FIGURE 5-7 Insulated siding *R*-value label.

FIGURE 5-8 Measuring spray foam.

Fiberglass and cellulose spray insulation in roof/ceiling thermal assemblies is measured with markers indicating each 1-inch thickness of installed insulation (Figure 5-9). The markers are attached to the trusses or joists and face the attic opening. **[Ref. C303.1.1.1]** This arrangement makes it easy for the contractor and building inspector to verify compliance without crawling through the attic with a tape measure.

FIGURE 5-9 Depth marker for blown-in attic insulation

FENESTRATION PRODUCTS

You Should Know

Testing in accordance with NFRC or AAMA/WDMA/CSA standards must be performed by an accredited and independent laboratory. ●

The glass, seals, frame and latching mechanisms in window, door or skylight assemblies are tested to produce the *U*-factor thermal performance rating, air leakage and light transmittance data. The results are printed on a label attached to each product. The label indicates the qualities of the fenestration, allowing the designer to specify properly and the builder and inspector to confirm compliance. Performance standards and labeling options are discussed below.

U-FACTOR & SHGC

The National Fenestration Rating Council (NFRC) develops testing criteria for whole-assembly performance for *U*-factors (NFRC 100) and solar heat gain coefficients (SHGC) (NFRC 200).

- The first and best option is to use a factory-affixed label. Not every window, door or skylight will have a label; some manufacturers don't participate in the testing program and some products will appear on a job site without a factory-affixed label due to a variety of reasons.
- If the assembly isn't factory labeled, the default tables in Section C303 must be used to determine the assumed values unless one of the alternatives below is viable.

- If the assembly has been tested and labeled per NFRC standards but the label is missing, NFRC has created a field labeling program that tracks the chain of custody from manufacture to delivery, thus providing a means to create a duplicate label and verify the performance of assembly. Field labeling is free of charge.
- Some fenestration, such as site-built storefront glazing never arrives with factory-affixed labels. For these, NFRC has developed a Component Modeling Approach (CMA) for use. This computerbased program will generate a site-specific certificate and label for each project. CMA is available from NFRC on a subscription basis. **[Ref. C303.1.3]**

FIGURE 5-10 Exposed foundation insulation protection.

AIR LEAKAGE

The NFRC and American Architectural Manufacturer's Association develop testing criteria for air infiltration (NFRC 400-2017 or AAMA/WDMA/CSA 101/I.S.2/A C440-17) to verify meeting the air leakage requirements in Table C402.5.2.

- The first and best option is to use a factory-affixed label. Not every window, door or skylight will have a label; some manufacturers don't participate in the testing program and some products will appear on a job site without a factory-affixed label due to a variety of reasons.
  - If the assembly has been tested and labeled per one of the standards but the label is missing, NFRC has created a field labeling program that tracks the chain of custody from manufacture to delivery, thus providing a means to create a duplicate label and verify the performance of assembly. Field labeling is free of charge.
  - Some fenestration, such as site-built storefront glazing, never arrives with factory-affixed labels. For these, NFRC has developed a component modeling approach (CMA) for use. This computer-based program will generate a site-specific certificate and label for each project. CMA is available from NFRC on a subscription basis.
  - Field-fabricated assemblies that are sealed in compliance with Section C402.5.2 are excepted from the labeling requirements. Please note that storefront glazing is NOT classified as field fabricated; refer to the definitions in Chapter 2 for additional clarification.
  - Fenestration in buildings that comply with the testing alternative of Section C402.5 are not required to meet the air leakage requirements in Table C402.5.2. **[Ref. C402.5.2]**

ADDITIONAL REQUIREMENTS

There are additional requirements for fenestration found throughout the IECC, including, but not limited to:

- Maximum area(s) **[Ref. C402.4.1]**
- Minimum skylight fenestration area **[Ref. C402.4.2]**
- Haze factor for skylights **[Ref. C402.4.2.2]**
- Provisions for dynamic glazing **[Ref. C402.4.3.3]**

EXPOSED FOUNDATION INSULATION

The code requires insulation on the exterior foundation to be protected from physical damage and direct sunlight. Basement and crawl space walls and slab-edge insulation must be covered with a rigid, opaque, weather-resistant material that covers the exposed material above grade and extends at least 6 inches below grade (Figure 5-10). **[Ref. C303.2.1]** This protection contributes to maintaining the thermal design performance of the installed insulation.

CHAPTER 6

Efficiency Requirements

The energy-efficient provisions for the building envelope, mechanical systems, service water heating and electrical power and lighting systems for new construction of commercial buildings are contained in Chapter 4 [CE] of the *International Energy Conservation Code*® (IECC®). Section C402 provides prescriptive requirements for building envelope code compliance. This section offers a simple guide in the form of *R*-value and *U*-factor tables, arranged by building envelope components, and answers the question "What do I have to do to comply?" The building envelope must be designed to meet the requirements of ASHRAE/ANSI/IESNA 90.1 or the IECC provisions found in either Table C402.1.3, Table C402.1.4 or the component performance alternative (Section C402.1.5). Mandatory provisions are also included for air leakage and air barriers, mechanical systems, pipe insulation and water heating and lighting systems.

Section C406 of the IECC requires one of eight choices of additional efficiency options when utilizing the prescriptive path of compliance for new construction. For code compliance, the designer or building owner must include either higher HVAC performance, efficient lighting systems beyond the mandatory requirements, enhanced lighting controls, a dedicated outdoor air system for certain HVAC equipment, high-efficiency service water heating or install an on-site renewable energy supply to satisfy this provision (see Figure 6-1). **[Ref. C401]**

These additional efficiencies for the prescriptive path are substantially matched by the reduction in allowable energy cost to 85 percent in the performance compliance path. **[Ref. C401.2, Option 3]**

Commissioning certain building mechanical and electrical power and lighting systems is required in Section C408. The commissioning agent will verify that the performance of these systems meets the criteria specified in the commissioning plan and the completion requirements of the design documents.

Typical construction techniques and materials used in high-rise office and apartment buildings and commercial businesses differ from those in typical low-rise residential buildings. It should be noted that most jurisdictions and state statutes require commercial architectural and engineering plan documents to be stamped and sealed by a licensed and registered professional. The following overview is limited to highlight the basic concepts of the commercial energy design provisions, given the complexity of commercial building design. Figure 6-2 provides a roadmap.

You Should Know

The specific location of insulation in the building envelope defines the conditioned space boundary. For example, attic insulation may be blown or laid in the uppermost ceiling or placed directly in the roof rafter spaces and under the roof sheathing. The first method describes an unconditioned attic and the second creates a conditioned attic space. ●

FIGURE 6-1 A hospital building with installed solar and wind-powered electric generation. (*Photo courtesy of Ken Hawkins*).

BUILDING ENVELOPE WALLS AND ROOFS

The building thermal envelope separates the inside environment from the outside environment, that is, the daily and seasonal exterior temperature swings and moisture events. The typical elements of the thermal envelope include roofs and above- and below-grade walls and floors, as listed in the prescriptive Tables C402.1.3 and C402.1.4. The key to establishing the boundaries of the thermal envelope is found in the definition of

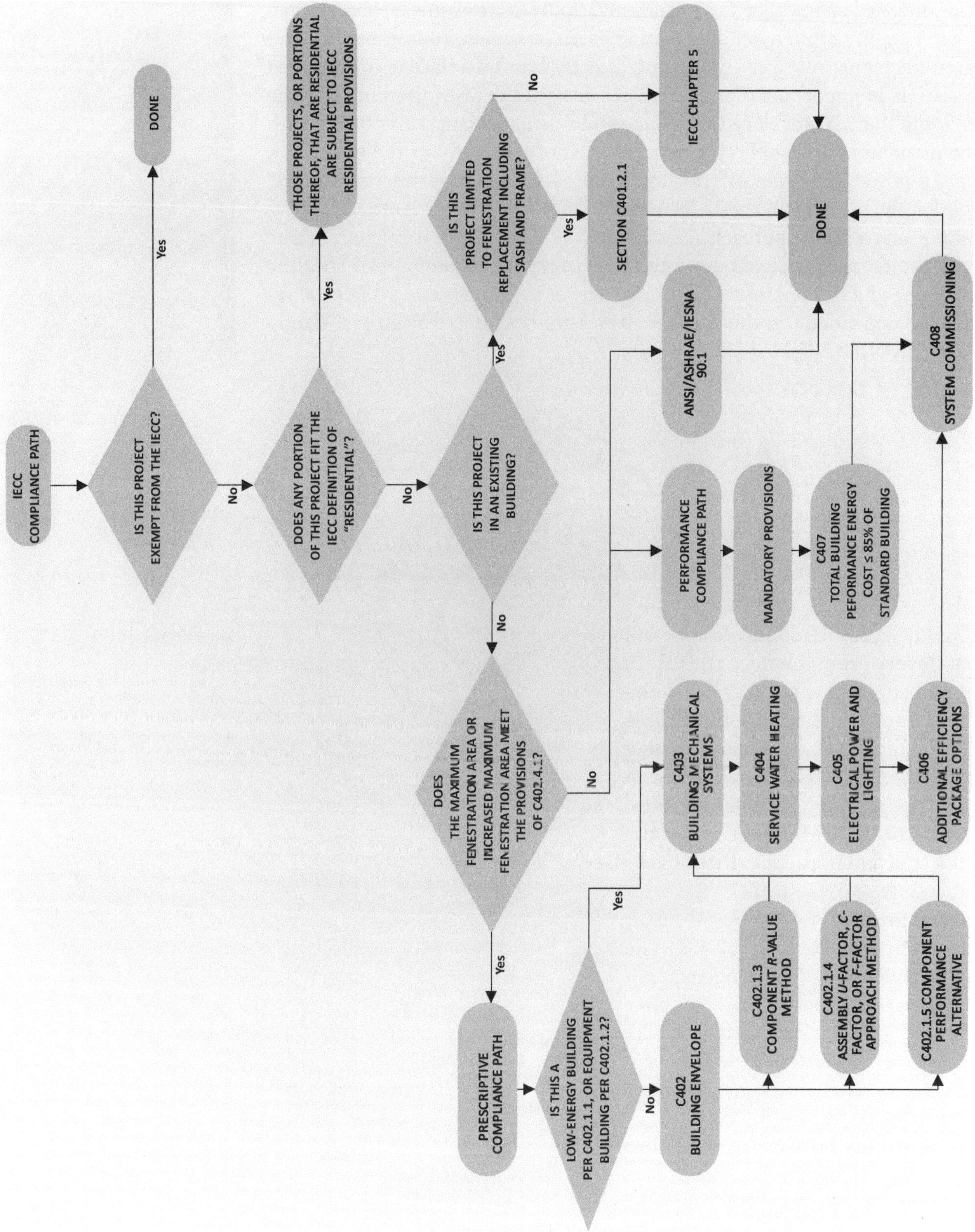

FIGURE 6-2 Flowchart

"Conditioned space." For the purpose of the IECC, *conditioned space* is "an area or room within a building being heated or cooled, containing uninsulated ducts, or with a fixed opening directly into an adjacent conditioned space." It is important that the project designer defines the limits of the building thermal envelope in the plans and demonstrates compliance with the mandatory and applicable requirements of Chapter 4 of the IECC.

As noted, the code offers three choices for prescriptive compliance: the *R*-value approach, the *U*-factor approach and the component performance alternative approach. **[Ref. C402.1]** Table C402.1.3 lists *R*-value requirements for individual components of a typical commercial building envelope and the required performance level by climate zones. The values are also dependent on the occupancy and are designated as either "Group R" or "all other" (Table 6-1).

Code Essentials

Continuous insulation (CI) can be placed on the inside or outside of an assembly and is continuous across all structural members with no significant thermal breaks (thermal bridges).

TABLE 6-1 *R*-value prescriptive requirements

Table C402.1.3	
Climate Zone 5	**Commercial building**
Insulation entirely above deck	R-30ci
Mass wall-above grade	R-11.4ci
Unheated slabs	R-10 for 24" below

Not every building design will use every component, and it is important to be familiar with the definitions and descriptions of these components. Only when building components are clearly identified in the drawings can the minimum thermal performance be specified for the roof, wall and floor types listed in the table (Figure 6-2). **[Ref C402.1.3]**

Table C402.1.4 is the *U*-factor alternative to Table C402.1.3. The applicable building envelope component is evaluated by an acceptable method, and the supporting calculations are submitted with the plan documents for review and approval (Figure 6-3). **[Ref C402.1.4]**

The component performance alternative method is described in its own section after the following discussion of fenestration products.

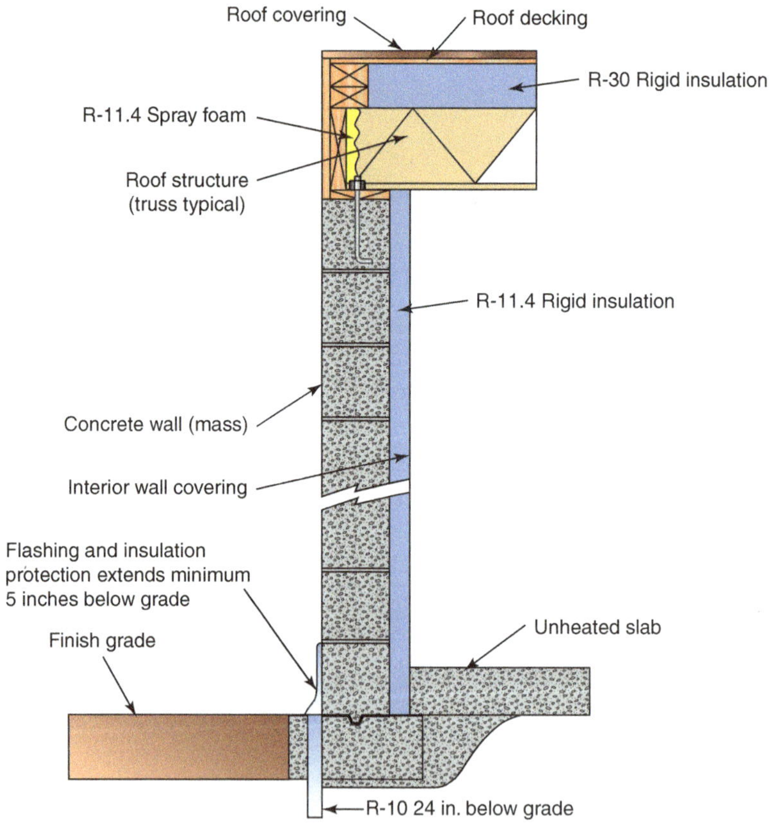

FIGURE 6-3 *R*-value insulation requirements for commercial building in Climate Zone 5.

WINDOWS, SKYLIGHTS AND DOORS

Table C402.4 of the IECC lists the prescriptive compliance requirements for commercial fenestration. The maximum window area allowed is 30 percent of the gross above-grade wall area. Skylight area is limited to not more than 3 percent. Certain exceptions are granted in areas with daylighting controls that allow the areas to be increased to 40 percent of above-grade walls and 5 percent of the roof area. The solar heat gain coefficient (SHGC) and projection factor are considered in establishing the requirements in this table. SHGC is an important consideration in energy efficiency and must be balanced with visible light transmittance in glazing design. The projection factor is a building element design strategy to provide shading to offset a portion of the heat gain through the fenestration. Dynamic glazing is recognized in the IECC and must be considered separately for compliance with the *U*-factor and SHGC requirements. **[Ref C402.4.3.3]**

COM*check* Software Version 4.3.5.1

Envelope Compliance Certificate

Project Information

Energy Code:	2015 IECC
Project Title:	2018 IECC Sample
Location:	Roscommon, Michigan
Climate Zone:	6a
Project Type:	New Construction
Vertical Glazing / Wall Area:	32%
Skylight / Roof Area	2%
Permit Date:	3/10/18
Permit No.	314159

Construction Site:	Owner/Agent:	Designer/Contractor:
507 Hitchcock Rd. Higgins Lake, Michigan	Jennifer C. Quali Worldwide Really Big Contractors and Cornhusk Doll Company 1313 Mockingbird Lane Otisville, Michigan 48523 BR 549 anythingforabuck@mail.com	Frank Hadley-Evers Wright Formby-Fore Functions Designs 951 Chicago Avenue Oak Park, Illinois 60302 900-837-5409 worldsgreatestarchitect@mail.com

Additional Efficiency Package(s)

Reduced interior lighting power. Requirements are implicitly enforced within interior lighting allowance calculations.

Building Area	Floor Area
1-Retail : Nonresidential	40000

Envelope Assemblies

Assembly	Gross Area or Perimeter	Cavity R-Value	Cont. R-Value	Proposed U-Factor	Budget U-Factor(a)
Roof: Insulation Entirely Above Deck, [Bldg. Use 1 - Retail]	40000	---	20.0	0.048	0.032
Skylight: Metal Frame with Thermal Break: Glass, with Curb, Perf. Specs.: Product ID 123, SHGC 0.35, [Bldg. Use 1 - Retail] (b)	1000	---	---	0.300	0.500
Floor: Heated Slab-On-Grade, Horizontal with vertical >= 4 ft., [Bldg. Use 1 - Retail] (c)	800	---	20.0	0.373	0.790
NORTH					
Ext. Wall: Concrete Block, 8in., Solid Grouted, Normal Density, Furring: Metal, [Bldg. Use 1 - Retail]	4000	0.0	17.0	0.050	0.080
Window: Metal Frame Curtain Wall/Storefront, Fixed, Perf. Specs.: Product ID 345, SHGC 0.20, PF 100.00, [Bldg. Use 1 - Retail] (b)	3000	---	---	0.300	0.360
Door: , Perf. Specs.: Product ID 234, SHGC 0.25, PF 50.00, [Bldg. Use 1 - Retail] (b)	264	---	---	0.300	0.770
EAST					
Ext. Wall: Concrete Block, 8in., Solid Grouted, Normal Density, Furring: Metal, [Bldg. Use 1 - Retail]	4000	0.0	17.0	0.050	0.080
Window: Metal Frame with Thermal Break: Fixed, Perf. Specs.: Product ID 456, SHGC 0.20, PF 50.00, [Bldg. Use 1 - Retail] (b)	500	---	---	0.300	0.360
SOUTH					

Project Title: 2018 IECC Sample Report date: 03/29/18
Data filename: Page 1 of 21

Assembly	Gross Area or Perimeter	Cavity R-Value	Cont. R-Value	Proposed U-Factor	Budget U-Factor(a)
Ext. Wall: Concrete Block, 8in., Solid Grouted, Normal Density, Furring: Metal, [Bldg. Use 1 - Retail]	4000	0.0	17.0	0.050	0.080
Window: Metal Frame with Thermal Break: Operable, Perf. Specs.: Product ID 345, SHGC 0.20, PF 50.00, [Bldg. Use 1 - Retail] (b)	500	---	---	0.300	0.430
Door: Insulated Metal, Swinging, [Bldg. Use 1 - Retail]	134	---	---	0.030	0.370
WEST					
Ext. Wall: Concrete Block, 8in., Solid Grouted, Normal Density, Furring: Metal, [Bldg. Use 1 - Retail]	4000	0.0	17.0	0.050	0.080
Window: Metal Frame Curtain Wall/Storefront, Fixed, Perf. Specs.: Product ID 345, SHGC 0.20, PF 50.00, [Bldg. Use 1 - Retail] (b)	800	---	---	0.300	0.360

(a) Budget U-factors are used for software baseline calculations ONLY, and are not code requirements.
(b) Fenestration product performance must be certified in accordance with NFRC and requires supporting documentation.
(c) Slab-On-Grade proposed and budget U-factors shown in table are F-factors.

Envelope PASSES: Design 0.5% better than code

Envelope Compliance Statement

Compliance Statement: The proposed envelope design represented in this document is consistent with the building plans, specifications, and other calculations submitted with this permit application. The proposed envelope systems have been designed to meet the 2015 IECC requirements in COM*check* Version 4.3.5.1 and to comply with any applicable mandatory requirements listed in the Inspection Checklist.

Name - Title Signature Date

Project Title: 2018 IECC Sample Report date: 03/29/18
Data filename: Page 2 of 21

FIGURE 6-4 *U*-value prescriptive compliance certificate.

COMPONENT PERFORMANCE ALTERNATIVE

The 2015 edition of the IECC inroduced a new building thermal envelope compliance path—the Component Performance Alternative. The formula allows various envelope components to be traded off against each other, provided that the overall calculated building heat loss of the proposed design is no greater than a code-compliant design. The five principal factors in the equation are described below. The idea is the same as the Total UA Alternative approach allowed as a residential compliance path in Section R402.1.5 but accounts for slab-edge *F*-factors, basement wall C-values and fenestration areas in excess of the code limits.

This optional path provides significant additional flexibility, allowing the designer to trade off the *U*-factors of various building envelope components without having to do a full Total Building Performance computation.

A greater window area might be acceptable, for example, with lower window *U*-values. Another example is the possible reduction of wall insulation in certain areas while roof insulation is increased. **[Ref. C402.1.5]**

Component Performance Alternative Formula

$A + B + C + D + E \leq \text{Zero}$ **(Equation 4-2)**

Where each factor represents the difference between the proposed design and a prescriptive design for

A. The sum of the *U*-factor for each envelope assembly times its area.
B. The sum of the *F*-factor for each slab-edge assembly times its length.
C. The sum of the *C*-value for each basement wall assembly times its area.
D. The additional amount for vertical glazing area in excess of maximum. (Substitutes the average wall *U*-factor for the average vertical glazing *U*-factor in the prescriptive case for the excess vertical glazing area.)
E. The additional amount for skylight area in excess of code maximum. (Substitutes the average roof *U*-factor for the average skylight *U*-factor in the prescriptive case for the excess skylight area.)

PERFORMANCE COMPLIANCE

The *total building performance* provisions of Section C407 of the IECC offer a fourth path for energy code compliance. This method uses computer software to predict the annual energy cost of the proposed design as compared to the standard design of the building components listed in Table C407.5.1(1). If the proposed building's annual energy cost is

You Should Know

The following factors are often used to perform a Total Building Performance:

- Orientation
- Fenestration Area
- Size of Overhangs
- Fenestration *U*-Factor
- Solar Heat Gain Coefficients and Daylighting Control Fraction (if automatic daylighting controls are used)
- *U*-factors of building envelope components
- Heat Capacity
- Insulation Position (for mass walls only) ●

> **You Should Know**
>
> When using Section 407 for the Total Building Performance compliance path, "The building energy cost shall be equal to or less than 85 percent of the standard reference design building."
> This is the ONLY location in the 2018 IECC where this often overlooked requirement is given." [Ref. C407.4.1] •

less than or equal to 85 percent of the standard design, the building complies with the code's goal of effective energy use. This 15-percent reduction produces similar savings as the Additional Efficiency Package Options (Section C406) with the prescriptive compliance options. Heating systems, cooling systems, service water heating systems, fan systems, lighting power, receptacle loads and process loads must be included in the calculations to determine the total building performance. [Ref C407.1]

This performance method allows trade-offs between the building envelope, HVAC system and lighting systems to achieve an overall building energy conservation result equal to or better than that of the other code compliance paths, although these individual component requirements in the modeling may be less than what is typically specified in the energy code tables. Regardless though, the mandatory requirements for components including air leakage performance details of the building envelope, HVAC system and lighting systems must be met. The compliance report must be submitted to the code official. [Ref C407.4.1]

LOW-ENERGY BUILDINGS

The IECC specifies three low-energy buildings that are exempt from the building thermal envelope provisions of the code:

1. Those with a peak design rate of energy usage less than 3.4 Btu/h • ft^2 (10.7 W/m^2) or 1.0 watt per square foot (10.7 W/m^2) of floor area for space conditioning purposes.
2. Those that do not contain *conditioned space*.
3. Greenhouses.

It should be emphasized that, along with the equipment building exemption noted below, the exemption is limited to building thermal envelope. All other regulated systems, if present, are required to comply with the code. [Ref. C402.1.1]

EQUIPMENT BUILDINGS

Separate electronic equipment buildings up to 500 square feet are exempt from the building thermal envelope provisions of the code when they meet a set of specified requirements. The thermostat must be restricted to a maximum of 50° F, and there is a maximum wall and roof *U*-factor that is dependent on climate zone. [Ref. C402.1.2]

COMMISSIONING

The following systems and their controls must be commissioned and completed in accordance with Section C408:

- Mechanical, service water heater, swimming pool water heating, spa water heating and lighting.

There are two exceptions:

1. Mechanical systems and service water heater systems in buildings where the total mechanical equipment capacity is less than 480,000 Btu/h (140.7 kW) cooling capacity and 600,000 Btu/h (175.8 kW) combined service water-heating and space-heating capacity.
2. Systems included in Section C403.5 that serve individual dwelling units and sleeping units. **[Ref. C408.2]**

Please note that Exception 1 applies exclusively to mechanical and water-heating systems, while any and all of the other systems—even if they are included in a project where the mechanical and service water heating are exempted—require commissioning per Section C408.

The design professional is required to provide construction document specifications and notes that clearly indicate provisions for commissioning and completion requirements. Evidence of mechanical systems commissioning and completion must be provided prior to the final mechanical and plumbing inspections. Preliminary and final commissioning reports shall be organized with mechanical and service hot water findings in separate sections to allow for independent review. **[Ref. C408.2.4]** An operating and maintenance manual must also be provided to the building owner for service hot water control systems, which covers maintenance and calibration information, including wiring diagrams, schematics and control sequence descriptions. **[Ref. C408.2.5]**

Requirements for lighting control system commissioning are based on the type of controller. Specifications are clearly spelled out for "occupant sensor controls," "time-switch controls" and "daylight responsive controls." Testing requirements of occupant sensor controls differ based on whether the project has more than seven sensors or seven and fewer. **[Ref. C408.3]**

Construction documents are required to specify the performance criteria of lighting, and like the requirements for mechanical systems, operation and maintenance manuals are now also required to be given to the building owners related to the operation, maintenance, inspection and recalibration of lighting controls. **[Ref. 408.2.5.2]**

CHAPTER

7 Controlling Air Leakage

Uncontrolled outdoor-to-indoor air exchange through the building envelope is known as *infiltration*. The opposite air movement, indoor-to-outdoor exchange, is *exfiltration*. Either type of air exchange affects energy costs. Conditioned air, heated or cooled, lost through exfiltration must be replaced by air from the outdoors. The building air-handling systems then use additional energy to bring the outdoor air to the desired indoor temperature. By the same process, hot or cold air that infiltrates the building places an increased burden on the system because the air leaking into the building must be conditioned and brought to the desired indoor temperature.

Air exchange is controlled through intentional ventilation. This may be as simple as opening a window or, as in most commercial buildings, through the HVAC system. Excess air exchange, controlled or uncontrolled, is contrary to the intent of the IECC. **[Ref. C101.3]**

THE AIR-BARRIER REQUIREMENTS

The air-leakage provisions of the IECC are mandatory for all compliance paths. Specific sections address air barriers and their construction, compliance requirements and options, penetrations, fenestration, air intakes and exhaust, loading dock seals and vestibules. **[Ref. C402.5]** The requirements cover provisions for a continuous air barrier to be designed, detailed and specified in the building plan set of documents. "*Continuous air barrier*" is defined in the IECC: "A combination of materials and assemblies that restrict or prevent the passage of air through the *building thermal envelope*." **[Ref. R202]** The air barrier may be inside or outside of the thermal envelope, and in all cases, must be continuous. An easy test to check for compliance is to trace the air-barrier location in the building sections and details on the construction documents. If this can be done without lifting the pencil off the paper, the air barrier is continuous.

Air-barrier materials cover and seal the joints and assemblies in the building thermal envelope that make up the air barrier. A single material or a combination of materials is allowed to satisfy the requirement. The materials, such as caulk or tape, must be securely installed along the entire length of the joint in the thermal envelope and perform during the pressure changes in the building.

You Should Know

ASTM E2178 is the testing standard to measure air permeance of air-barrier materials. Liquid-applied membranes, mechanically fastened commercial building wraps and rigid panel materials are common air-barrier materials in commercial construction. The test results of the specific materials are reported and must comply with the IECC. ●

Compliance

Three methods are available to the designer and builder to comply with the air-barrier provisions for the solid portions of the building envelope:

(1) compliance with a list of 17 prescriptive materials with a specific air permeability,

(2) acceptance of assemblies with a specified average air-leakage rate, or

(3) performance of a test of the completed buildings to verify the actual air-leakage rate. **[Ref. C402.5.1.2]**

In the first option, 17 common building materials are specifically named (Table 7-1). When any of these materials is installed according to the manufacturer's instructions and the joints are properly sealed, this requirement is met for that portion of the building. Any material that meets the specified air-leakage rate not greater than 0.04 cfm/ft^2 at a differential of 0.3-inch water gauge (75 Pa) per ASTM E2178 also complies. **[Ref. C402.5.1.2.1]**

The second option relies on the performance of the assemblies versus the individual component materials that were listed above. Four assemblies are specifically listed that are deemed to meet the required limits (Table 7-2). The referenced tests apply to curtain walls, air-barrier systems and large window assemblies. **[Ref. C402.5.1.2.2]**

FIGURE 7-1 Penetrations in the air barrier must be sealed.

TABLE 7-1 Common building materials that comply with the air-leakage requirements

Air-barrier material (prescriptive)	Thickness (minimum)
Plywood	$^3/_8$ in. (10 mm)
Oriented strand board	$^3/_8$ in. (10 mm)
Extruded polystyrene insulation board	$^1/_2$ in. (12.7 mm)
Foil-back polyisocyanurate insulation board	$^1/_2$ in. (12.7 mm)
Closed-cell spray foam minimum density of 1.5 pcf	1-$^1/_2$ in. (38 mm)
Open-cell spray foam density between 0.4 and 1.5 pcf	4.5 in. (113 mm)
Exterior or interior gypsum board	$^1/_2$ in. (12.7 mm)
Cement board	$^1/_2$ in. (12.7 mm)
Built-up roofing membrane	Thickness not applicable
Modified bituminous roof membrane	Thickness not applicable
Fully adhered single-ply roof membrane	Thickness not applicable
A Portland cement/sand parge or gypsum plaster	$^5/_8$ in. (15.9 mm)
Cast-in-place and precast concrete	Thickness not applicable
Sheet steel or aluminum	Thickness not applicable
Fully grouted concrete block masonry	Thickness not applicable
Solid or hollow masonry constructed of clay or shale masonry units	Thickness not applicable
Any material with air-leakage rate ≤0.004 cfm/ft^2 at 0.3-inch water gauge (75 Pa) per ASTM E2178	Thickness adequate to meet the test requirements

TABLE 7-2 Assemblies deemed to comply with the air-leakage requirements.

Material(s)	Construction specific requirements
Concrete masonry walls	Coated with one application of block filler. Coated with two applications of paint or sealer coating.
Masonry walls constructed of clay or shale masonry units	Nominal width of 4 inches (102 mm) or more.
Portland cement/sand parge or gypsum plaster	Minimum ½ inch (12.7 mm) thickness.
Any assembly with an average air leakage rate ≤0.04 cfm/ft^2 at 0.3-inch water gauge (75 Pa) per ASTM E2357, or ASTM E1677 or ASTM E283.	Thickness adequate to meet the infiltration test requirements.

Pressure testing of the completed building is the third option to demonstrate air-barrier or air-leakage compliance. It should be noted that the mere presence of compliant materials and/or assemblies found in the first two options above may not be adequate to create a continuous air barrier and thus preclude the requirement for pressure testing of the completed building. Pressure testing of the building is also an alternate option to demonstrate fenestration air leakage when assemblies are not tested and labeled (see Figure 7-2) **[Ref. C402.5.2]**

Penetrations

In every building there are numerous penetrations, connections and joints comprised of both similar and dissimilar materials—each providing options for infiltration and/or exfiltration if not properly detailed and installed. Electrical boxes, mechanical vents and hose bibbs are common penetrations that must be gasketed or caulked (Figure 7-1). Joints between materials create a break in the air barrier and must be caulked, taped or otherwise sealed. The sealing and air-barrier materials must be compatible and installed

FIGURE 7-2 Assembly of commercial blower for air leakage testing.

according to the manufacturer's instructions. It is incumbent on the code official to demand air sealing details [Ref. C103.3], the design professional(s) to provide them adequate for review and the inspector to verify the installation. If these details are inadequate, the building is required to be pressure tested, regardless of the presence of materials or assemblies that are deemed to be compliant. [Ref. C402.5.1.1]

INFILTRATION RATES

Fenestration components are listed in Table C402.5.2 of the IECC and grouped by the applicable test procedure. Each fenestration component fills a hole in the opaque (solid) exterior building wall thermal envelope and air barrier. It is important to identify the windows, skylights and most doors on the plans and schedules. Air infiltration performance should be detailed in the construction documents. The National Fenestration Rating Council has developed a Component Modeling Approach (CMA), which can produce a certificate and label for site-built glazing in compliance with the construction documents provisions. [Ref. C103.2] Field-fabricated fenestration that is sealed in compliance with Section C402.5.1 is not required to be tested. It is worthwhile to remember the differing definitions of "Field fabricated" and "Site built"—remembering that storefront assemblies are required to be tested—either as an assembly or as a part of the entire building. Field inspection will confirm that the units are listed, labeled and installed according to the manufacturer's instructions. Service-access openings to shafts and chutes and doors to stairways and elevator lobbies from conditioned space must be listed and gasketed, weather-stripped or sealed.

Code Essentials

Section 716 of the *International Building Code* (IBC) is referenced in an exception to the IECC air-leakage provisions. IBC Chapter 7 is titled "Fire and Smoke Protection Features," and Section 716 is "Opening Protectives." Underwriters Laboratory (UL) 1784 is a standard for "Air Leakage Tests of Door Assemblies and Other Opening Protectives." An important distinction is made here in that the UL 1784 requirements apply to doors (fenestration) in the continuous air barrier that are required to be fire and smoke protected based on the building code. The doors required to maintain fire and smoke control must be identified on the plans and schedule to comply with UL 1784. **[Ref. C402.5.4, exceptions]** ●

OUTDOOR AIR OPENINGS

Mechanical ventilation systems bring fresh air into the building interior and exhaust stale air to the outdoors. The openings must penetrate the air barrier to get the job done. Uncontrolled airflow through the outdoor air vents and in stairway enclosures and elevator shaft vents is accelerated by pressure imbalances between the building interior and exterior. Motorized dampers are required in these openings to control the air leakage through the air barrier (Figure 7-3). The dampers must be listed and tested.

Gravity dampers comply in buildings less than three stories above grade; for ventilation air intake, exhaust and relief dampers in any building in Climate Zones 1, 2 and 3; and in buildings where the design exhaust capacity is not greater than 300 cfm. **[Ref. C403.7.7 Exceptions 1, 2 and 3]**

These particular exceptions are a good reminder to read and reread the code provisions completely, as they are an example of an opportunity to save significant time and money and still comply with the applicable energy code provisions for dampers.

LOADING DOCKS

Most commercial buildings require loading docks for the delivery of food, dry goods and supplies. The exterior of the loading dock door perimeter must have a weatherseal installed to minimize airflow (Figure 7-4). The provision applies to all loading dock doors in all climate zones. Noncompliant loading dock doors are pictured in Figure 7-5. **[Ref. C402.5.6]**

FIGURE 7-3 Motorized dampers control the air leakage through the air barrier.

Code Essentials

The "Building thermal envelope" is defined as the "basement walls, exterior walls, floors, ceilings, roofs and any other building element assemblies that enclose conditioned space or provide a boundary between conditioned space and exempt or unconditioned space." [Ref C202] •

VESTIBULES

The common architectural meaning for the term "*vestibule*" is "a passage, hall or antechamber between the outer door and the interior parts of a house or building." This transition space, from outdoors to indoors, is a break in the air barrier and a direct path of infiltration every time the door is used. Vestibule provisions include self-closing doors and designs that allow closing one door behind you before opening the next. Although revolving doors do not require vestibules, any other entrance door adjacent to the revolving door does require a vestibule. Having a revolving door at the building entrance does not satisfy the intent of the vestibule requirement for these other doors (Figure 7-6). **[Ref. C402.5.7]**

In most buildings, vestibules designed to meet the energy code provisions result in significant energy savings. The IECC recognizes that some factors—such as the size of a compliant design, the use of certain exterior doors and locations in warm climates—may create exemptions from the vestibule requirements.

Vestibules are not required for the following:

1. Buildings in Climate Zones 1 and 2.
2. Doors not intended to be used by the public, such as doors to mechanical or electrical equipment rooms, or intended solely for employee use.
3. Doors opening directly from a sleeping unit or dwelling unit.
4. Doors that open directly from a space less than 3,000 square feet (298 m^2) in area.
5. Revolving doors.

FIGURE 7-4 Compliant loading dock seal.

FIGURE 7-5 Noncompliant loading dock.

6. Doors used primarily to facilitate vehicular movement or material handling and adjacent personnel doors.
7. Doors that have an air curtain with a velocity of not less than 6.56 feet per second (2 m/s) at the floor that have been tested in accordance with ANSI/AMCA 220 and installed in accordance with the manufacturer's instructions. Manual or automatic controls shall be provided that will operate the air curtain with the opening and closing of the door. Air curtains and their controls shall comply with Section C408.2.3 (Figure 7-7).

Accessibility must also be considered when designing the distance between doors. The building code and accessibility laws require certain minimum sizes for vestibules (Figure 7-8).

You Should Know

Although revolving doors do not require vestibules, other doors adjacent to revolving doors will still need to meet the vestibule requirements. This applies to doors provided for accessibility and/or required means of egress. ●

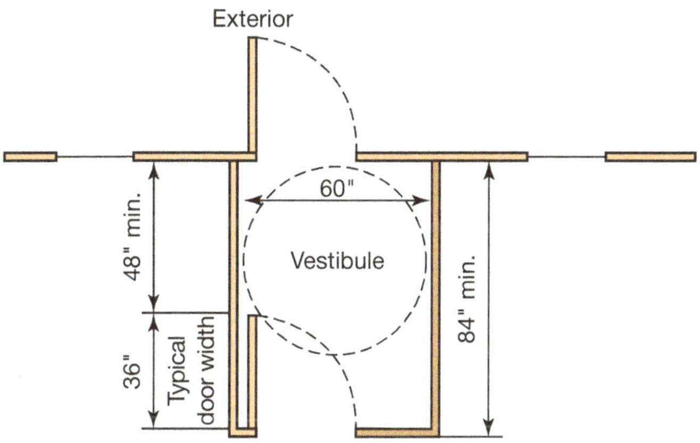

FIGURE 7-8 Typical minimum dimensions for a compliant vestibule.

FIGURE 7-6 Revolving doors do not require vestibules.

FIGURE 7-7 Air curtain on entrance door (*Photo Courtesy of Berner International).*

CHAPTER

8 Mechanical Systems

When paying a monthly energy bill, we all pay attention to how much it costs to heat or cool our building. It stands to reason why a code devoted to saving energy would do the same thing. Much of the *International Energy Conservation Code®* (IECC®) is focused on taking what other codes require, such as mechanical, building, plumbing and electrical and finding ways to lower the effects on the building's mechanical systems while not interfering with the health and life safety of the building's occupants. In order to have a mechanical system run efficiently, we need to build a good thermal envelope that separates the air we are paying to heat and cool from the air that we are not paying for. We also need to utilize that envelope to offset some of the lighting requirements where possible and design a mechanical system that will not require a building owner to choose between paying the mortgage or the energy bill. Mechanical systems produce and circulate tempered and treated air and hydronic fluids to make people comfortable inside a building. The energy efficiency of the heating, ventilation and air conditioning (HVAC) system depends on it operating only when needed, being off when it is not needed and using no more energy than necessary while it is operating. To accomplish this, the mandatory provisions in the IECC [Ref. C403.2] include criteria for sizing the HVAC equipment, minimum equipment efficiency, system operational controls, ventilation volume controls, energy recovery ventilation (ERV) systems, air duct insulation and sealing, piping insulation, guestroom HVAC system controls, refrigeration equipment performance and controlling systems outside a building. Provisions specific to more complex systems such as variable air volume and multiple-zone systems [Ref. C403.6.1] are clearly divided into separate, additional code requirements. It is important to correctly classify the system to determine the specific code provisions in addition to the mandatory requirements for all HVAC systems.

Code Essentials

The code specifies the design temperatures so the equipment can be properly sized. Design temperatures are based on climate data within the larger climate zones referenced in the IECC. The information allows the HVAC to be properly sized, allowing efficient operation. Equipment designed to operate over the design temperature for heating or below the design temperature for cooling defeats the energy-efficient operation and design intent of the IECC. The indoor design temperature input is limited to 72°F for heating and 75°F for cooling as required by Section C302 ●

THE MANDATORY REQUIREMENTS

Equipment sizing

Accurate load calculations are essential and the first step in mechanical system design. The design load calculation provides the information required to determine the right size of HVAC equipment for the building. The output rating of the heating and cooling equipment listed in the mechanical schedule must not be greater than that of the smallest available equipment size that exceeds the calculated heating and cooling loads. [Ref. C403.3] Both undersized or oversized HVAC equipment is inefficient because the run cycles are either too long (undersized) or too short (oversized).

Equipment efficiency

Heating and cooling equipment must meet the minimum efficiency requirements in Tables C403.3.2(1–9) of the IECC. The tables are grouped to help the user locate equipment type and equipment size. The detailed descriptions and size choices offered in the tables guide the designer and plans examiner in determining the minimum efficiency requirements specific to the equipment in the drawings and schedules (Table 8-1).

The efficiency of air-conditioning equipment is rated by several metrics established in the referenced test procedure. The Air-Conditioning, Heating, and Refrigeration Institute (AHRI) offers a simple and understandable basis for establishing the standards used to compare this type of equipment. The seasonal energy efficiency ratio (SEER) is the ratio of the seasonal cooling output divided by the total electric energy input. The energy efficiency rating (EER) is the cooling output at a particular operating point divided by the electrical power input. The integrated

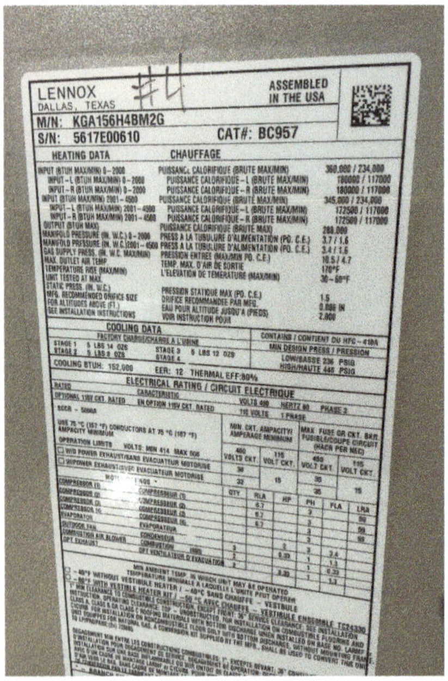

FIGURE 8-1 Output rating.

energy efficient ratio (IEER) measures the part-load performance of the unit.[1]

Heating equipment efficiency is rated similarly to cooling equipment, using test metrics and standards. The U.S. Department of Energy (DOE) explains annual fuel utilization efficiency (AFUE) as "the ratio of heat output of the furnace or boiler compared to the total energy consumed by the furnace or boiler."[2] The coefficient of performance (COP) is the ratio of change in units of heat or cooling energy output by the equipment divided by the energy used for the output (Figure 8-1). A heat pump COP of 3.3 means 3.3 units of heat energy, expressed in kilowatt-hours (3.3 kWh), are provided for every 1 unit of energy (1 kWh) used. The heating season performance factor (HSPF) is related to the COP for air source heat pumps, expressed in units of BTU/Wh.

Equipment system control

Each heating and cooling zone must be independently controlled by a programmable thermostat. A "*Zone*" is defined as "A space or group of spaces within a building with heating or cooling requirements that are sufficiently similar so that desired conditions can be maintained throughout using a single control device." **[Ref. C202]**

TABLE 8-1 Heating and cooling minimum efficiency requirements

Table C403.3.2(2) Minimum efficiency requirements: electrically operated unitary and applied heat pumps

Equipment type	Size category	Heating section type	Subcategory or rating condition	Minimum efficiency	Test procedure
Air cooled (cooling mode)	< 65,000 Btu/h	All	Split system	14.0 SEER	AHRI 210/240
			Single packaged	14.0 SEER	
Through-the-wall, air cooled	< 30,000 Btu/h	All	Split system	12.0 SEER	
			Single packaged	12.0 SEER	
Single-duct high-velocity air cooled	< 65,000 Btu/h	All	Split system	11.0 SEER	
Air cooled (cooling mode)	≥65,000 Btu/h and < 135,000 Btu/h	Electric resistance (or None)	Split system and single package	11.0 EER 12.0 IEER	AHRI 340/360
		All other	Split system and single package	10.8 EER 11.8 IEER	
	≥135,000 Btu/h and < 240,000 Btu/h	Electric resistance (or None)	Split system and single package	10.6 EER 11.6 IEER	
		All other	Split system and single package	10.4 EER 11.4 IEER	
	≥240,000 Btu/h	Electric resistance (or None)	Split system and single package	9.5 EER 10.6 IEER	
		All other	Split system and single package	9.3 EER 9.4 IEER	

[1] ANSI/AHRI Standard 210/240 with Addenda 1 and 2 (formerly ARI Standard 210/240), *2008 Standard for Performance Rating of Unitary Air-Conditioning & Air-Source Heat Pump Equipment.*
[2] U.S. Department of Energy, *Energy Efficiency & Renewable Energy*, "Understanding the Efficiency Rating of Furnaces and Boilers."

The control device must be configured to set back the HVAC temperature demand when the building is not in use, and must be configured to start the system to make the building comfortable when the building is occupied. [Ref. C403.4.2]

Large buildings with HVAC systems that serve zones over 25,000 square feet in floor area or that span more than one floor must be separated into isolation areas if those areas are not always operating or occupied at the same time so that if one area isn't being operated or occupied, it isn't being heated or cooled. Each isolation area must be controlled independently with appropriate isolation devices and controls. [Ref. C403.2.1]

Snow and ice melt system controls

Snow and ice melt systems linked to the building's energy system, either hydronic (Figure 8-2) or electric, must have automatic controls. The controls monitor temperature and precipitation. When the pavement and air temperatures are safely above the point beyond that snow will not stick and ice will not accumulate, the system energy use is greatly reduced. The controls allow the system to idle in the off mode. This lessens the thermal shock effect and prolongs the functional life of the pavement slab. A manual control is allowed so that the system can be shut off completely during warm weather (Figure 8-3). [Ref. C403.12.2]

You Should Know

In the 2018 IECC, if a vestibule is heated, it must have controls configured to shut off the heating source when the outside air is greater than 45°F. There also must be a thermostat in heated and/or cooled vestibules that limits heating to not greater than 60°F and cooling to not less than 85°F. [Ref C403.4.1.4] ●

FIGURE 8-2 Snow and ice melt systems.

FIGURE 8-3 Manual control is allowed for complete shutdown.

FIGURE 8-4 Heat-traced process piping (*Photo courtesy of Pentair*).

HEAT-TRACED PIPING

In severe cold winter climates, freeze protection systems are often put in place to protect outdoor piping and heat exchangers (Figure 8-4). This strategy requires that the pipe-heating element have automatic controls to turn off the system when the outdoor air temperature is above 40°F or when it is determined the pipe fluid will not freeze. [Ref. C403.12.3] The requirements for heat-traced piping are similar to those for snow-melt systems in intent and effect. There is no purpose in expending energy to power a system combatting the negative effects of freezing temperatures when those temperatures cannot occur.

VENTILATION

The *International Mechanical Code*® (IMC®) requires ventilation in all buildings. Natural or mechanical ventilation is acceptable as long as the minimum outdoor air supply complies with the IMC. Many HVAC systems combine natural and mechanical ventilation. The natural ventilation strategy offers significant opportunities for energy savings. Mechanical ventilation controls are programmed to maximize the available natural ventilation. This is an efficient approach to space conditioning and indoor air quality. [Ref. C403.2.2 and C403.7]

Demand-controlled ventilation (DCV)

Demand-controlled ventilation (DCV) varies the minimum outdoor air supply to larger assembly areas in the building. DCV systems bring in fresh outdoor air when sensors in the monitored building area measure increased levels of carbon dioxide. Carbon dioxide will accumulate in a room when it is occupied, and the ventilation system may not replace the stale air at a rate to prevent the buildup. The DCV adjusts the air supply based on the number of people in the space. If used for a small function, the system will not fully open and will provide a reduced level of outdoor air while a large function would result in the system opening up and operating near or at its designed capacity. Every zone more than 500 square feet in area with an average occupant load of 25 people or more per 1,000 square feet must have a DCV system installed when the HVAC system has an air-side economizer, automatic modulating control of the outdoor air damper or a design outdoor airflow greater than 3,000 cfm. There are exceptions to this provision allowing for specific system components, designs and controls. [Ref. C403.7.1]

You Should Know

According to the IMC, the purpose of building ventilation is to protect the health of building occupants by controlling indoor air quality. Outdoor pollution may be introduced to the indoor environment, and indoor contaminants may be produced by equipment, furnishings and processes. Ventilation is necessary to control these typical air contaminants. Properly designed and functioning ventilation systems are necessary and help create a healthy and comfortable environment for building occupants.●

Energy recovery ventilation (ERV)

Energy recovery ventilation (ERV) is an exchange process that transfers the warm or cool temperature in exhaust air and exchanges the energy to temper the supply air coming into the building. Tables C403.7.4 (1 and 2) of the IECC list climate zones and percentage of supply airflow rates triggering the ERV requirement. The tables are divided based on whether the ventilation system will be operating less than 8,000 hours per year. Eleven specific exceptions qualify conditions where ERV systems are not required, and include heating energy recovery in Climate Zones 1 and 2; cooling energy recovery in Climate Zones 3C, 4C, 5B, 5C, 6B, 7 and 8; for systems serving spaces designed to stay cool and heated to less than 60°F; and for systems operating less than 20 hours a week. **[Ref. C403.7.4]**

FIGURE 8-5 ERVs transfer moisture along with heat.

ERVs differ from HRVs because they transfer a controlled amount of moisture along with the heat (Figure 8-5). This helps control the humidity in the building. In the summer, an ERV will remove some of the more humid supply air to the exhaust air. In the winter, the ERV will add some moisture to the supply air by removing it from the exhaust air. This transfer is accomplished by an enthalpy wheel.

Enthalpy in this context is the exchange of heat *and* moisture. It takes a certain amount of heat energy to change water to steam, just as it takes a certain amount of energy to add water vapor (humidity) to, or remove it from, the interior air. This moisture transfer, unique to an ERV, is the defining quality of an enthalpy wheel. The U.S. Department of Energy estimates that ERVs are 70–80 percent efficient in recovering energy in the exhaust air and adding it to the supply air.

You Should Know

A heat recovery ventilator (HRV) is one type of energy recovery system. It only captures heat from the exhaust air and transfers some of the heat to the supply air. These energy recovery systems are common in cool, dry climates.

DUCTS AND PLENUMS

All air ducts and plenums must be insulated to minimize heat loss and sealed to minimize air leakage. R-6 insulation is the minimum requirement for ducts and plenums in unconditioned space (Figure 8-6). When a duct or plenum is outside the building, R-8 insulation is the minimum required in Climate Zones 1-4 and R-12 in Climate Zones 5-8. The energy code provisions specific to sealing ducts, joints and seams in ductwork and HVAC components reference IMC Section 603.9. Compliant methods include listed and labeled tapes and mastics and continuously welded and locking joints and seams (Figure 8-7). Ducts, duct fittings, dampers, plenums and fans must be sealed as required in the energy and mechanical codes. Specific construction requirements for low-pressure duct systems (operating at less than or equal to 2 inches of water gauge pressure), medium-pressure duct systems (more than 2 inches but less than 3 inches of water

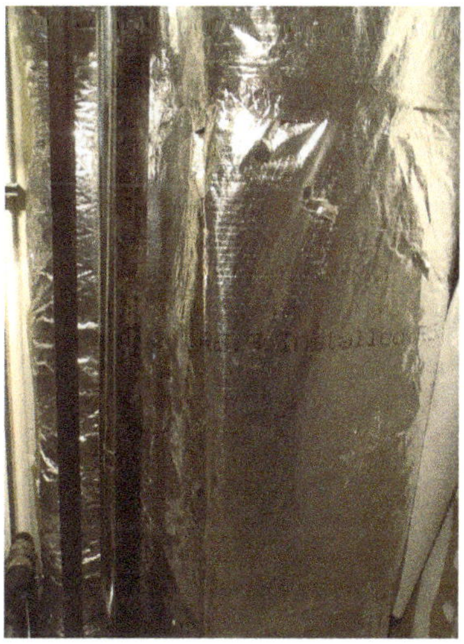

FIGURE 8-6 R-6 is the minimum requirement for ducts and plenums in unconditioned space.

FIGURE 8-7 Continuously welded and locked joints and seams.

gauge pressure) and high-pressure duct systems (more than 3 inches of water gauge pressure) are detailed in this section. **[Ref. C403.11]**

Open combustion air ducts

In Climate Zones 3-8, when fuel-burning appliances are supplied by open combustion air ducts, the rooms containing the appliances and ducts supplying them must be separated from conditioned spaces in compliance with the building thermal envelope provisions of the code. If the combustion air duct passes through conditioned space, it must be insulated to a minimum of R-8. Exceptions to this requirement include certain fireplaces and stoves. **[Ref. C402.5.3]**

Pipe insulation

Pipes carrying fluids serving HVAC systems must be insulated according to Table C403.11.3 of the IECC. The hotter the fluid moving through the pipe, the more insulation is required. This is partially due to the fact that the cooling loads in commercial buildings are usually greater than the heating loads. The heat coming off the pipes carrying very hot fluids unnecessarily adds to the heating load if the pipes are not properly insulated (Table 8-2). Note that the IECC table does not require insulation for pipes moving fluids that operate between 60°F and 105°F. This is restated in Exception 3. **[Ref. C403.11.3]**

Pipe insulation must be protected when exposed to sunlight, moisture, wind and possible damage during HVAC equipment maintenance (Figure 8-8) **[Ref. C403.11.3.1]**.

Commissioning and completion

Buildings with a total HVAC cooling equipment capacity equal to or greater than 480,000 Btu/h (commonly expressed as 40 tons of cooling) or total heating and water heating equipment capacity equal to or greater than 600,000 Btu/h must meet the specific requirements in Section C408 of the IECC. **[Ref. C408.1]** The provisions include a commissioning plan, a preliminary and a final commissioning report, adjusting and balancing the system, and testing and documentation, including a mandatory commissioning compliance checklist **[Ref. Figure C408.2.4]**.

You Should Know

The U.S. Environmental Protection Agency (EPA) estimates that 20 percent of the energy efficiency of heating and cooling systems can be lost because of duct air leaks outside of the conditioned space. Measurable gains in energy efficiency can be made by ensuring ducts, plenums and HVAC components are sealed as required by the provisions of the IECC and IMC. Building science shows that duct leakage also affects the positive and negative pressures in a building, playing a role in air and contaminant movement in and through the building assemblies. ●

TABLE 8-2 Table C403.11.3: Minimum pipe insulation thickness (thickness in inches)[a, c]

Fluid operating temperature range and usage (°F)	Insulation conductivity		Nominal pipe or tube size (inches)				
	Btu • in./(h • ft2 • °F)[b]	Mean Rating Temperature, °F	< 1	1 to < 1½	1½ to < 4	4 to < 8	≥ 8
> 350	0.32 – 0.34	250	4.5	5.0	5.0	5.0	5.0
251–350	0.29 – 0.32	200	3.0	4.0	4.5	4.5	4.5
201–250	0.27 – 0.30	150	2.5	2.5	2.5	3.0	3.0
141–200	0.25 – 0.29	125	1.5	1.5	2.0	2.0	2.0
105–140	0.21 – 0.28	100	1.0	1.0	1.5	1.5	1.5
40–60	0.21 – 0.27	75	0.5	0.5	1.0	1.0	1.0
< 40	0.20 – 0.26	50	0.5	1.0	1.0	1.0	1.5

For SI: 1 inch = 25.4 mm, °C = [(°F) - 32]/1.8.

a. For piping smaller than $1^1/_2$ inches and located in partitions within conditioned spaces, reduction of these thicknesses by 1 inch shall be permitted (before thickness adjustment required in footnote b) but not to a thickness less than 1 inch.

b. For insulation outside the stated conductivity range, the minimum thickness (T) shall be determined as follows:

$$T = r[(1 + t/r)^{K/k} - 1]$$

where:

T = minimum insulation thickness,

r = actual outside radius of pipe,

t = insulation thickness listed in the table for applicable fluid temperature and pipe size,

K = conductivity of alternate material at mean rating temperature indicated for the applicable fluid temperature (Btu • in/h • ft² • °F) and

k = the upper value of the conductivity range listed in the table for the applicable fluid temperature.

c. For direct-buried heating and hot water system piping, reduction of these thicknesses by $1^1/_2$ inches shall be permitted (before thickness adjustment required in footnote b but not to thicknesses less than 1 inch.

Outdoor heating

Radiant heat systems are required when outdoor spaces are heated (Figure 8-9). An occupancy sensor or timer switch is required so that the heating system turns off when no people are in the space. [Ref. C403.12.1]

HVAC SYSTEMS SERVING GUESTROOMS

Hotel and motels that have more than 50 guestrooms must have controls in each room that are configured to automatically raise the cooling setpoint and lower the heating setpoint by not less than 4°F from the occupant setpoint within 30 minutes of the occupant leaving the room. The control must also be configured to raise the cooling setpoint to at least 80° and lower the heating setpoint to not less than 60°F when the room is unrented or has not been continuously occupied for at least

FIGURE 8-8 Pipe insulation must be protected.

FIGURE 8-9 Occupancy sensor or time switch required for outdoor radiant heating.

FIGURE 8-10 Card key controls. *(Courtesy of New Buildings Institute)*

16 hours. Card key controls are considered to comply with these requirements. Networked guestroom control systems that indicate when a room is unoccupied for more than 30 minutes also are considered compliant controls. **[Ref. C403.7.6]**

Controls or isolation devices are also to be provided in guestrooms that automatically turn off the ventilation and exhaust fans within 30 minutes of occupants leaving the room. **[Ref. C403.7.6.2]**

ECONOMIZERS

While economizers are considered to be a prescriptive code requirement, they are allowed to be traded off when using the performance approach to compliance. In any case, they are considered to play an important role in the efficiency of most cooling systems. Economizers provide cool outside air to indoor space without the energy cost of producing conditioned air. This is essentially "free cooling" managed by appropriate fan and damper controls and temperature sensors. Economizers are not appropriate in all climates or cost effective for all systems, and specific exceptions for efficiency, hours of weekly operation and design parameters detail when economizers are not required. **[Ref. C403.5]**

Required economizers must be sized to provide all of the indoor cool air from the outside and must include a relief air outlet to prevent building overpressurization. **[Ref. C403.5.3]** Certain economizers must also be equipped with a fault detection and diagnostic system to ensure proper operation of this energy saving device **[Ref. C403.5.5]**

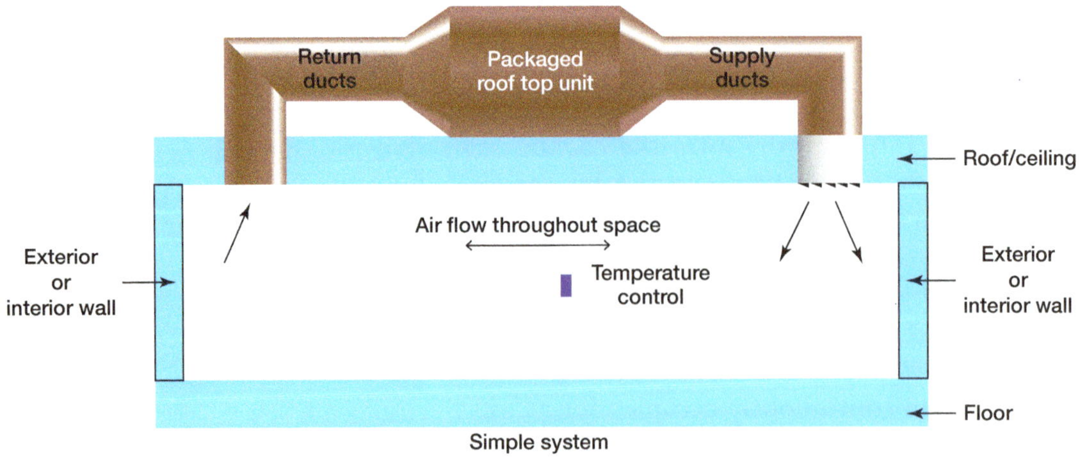

FIGURE 8-11 Simple HVAC system. *(Courtesy of Pacific Northwest National Laboratory)*

TABLE 8-3 Economizer requirements

Climate zones	Economizer requirement
1A, 1B	No requirement
2A, 2B, 3A, 3B, 3C, 4A, 4B, 4C, 5A, 5B, 5C, 6A, 6B, 7, 8	Economizers on all cooling systems ≥ 54,000 Btu/h[a]

For SI: 1 British thermal unit per hour = 0.2931 W.

a. The total capacity of all systems without economizers shall not exceed 300,000 Btu/h per building, or 20 percent of its air economizer capacity, whichever is greater.

FAN MOTORS

Fan motors are an integral component of the HVAC system. They move large volumes of air through ducts and plenums in the building ventilation system. HVAC fans with motors rated 7.5 horsepower or greater must have the capability to operate at two-thirds of full speed or less and have controls that automatically change the fan speed to control fluid and condensing temperatures. [Ref. C403.8.4] The power required for a fan is proportional to the cube of the fan's speed, so a one-third reduction in fan speed results in a 70-percent reduction in energy consumption.

Table C403.8.1(1) provides two limitation options for fans as a function of the maximum design supply airflow rate. Option 2, the maximum brake horsepower, provides pressure drop adjustments for systems with filters, energy recovery devices and others. These adjustments are listed in Table C403.8.1(2). Fans over the 5-horsepower threshold must also be listed with a fan efficiency grade (FEG) of not less than 67. Compliance for this provision is based on the manufacturer's certified data.

You Should Know

Economizers save the most energy when they are enthalpy controlled. Such a control also accounts for the outdoor air humidity in addition to temperature.●

Mechanical systems serving multiple zones

Provisions for efficient operation of more complex HVAC systems serving multiple zones require controlled variable air volume (VAV) fan systems configured to deliver tempered air only to the zone needing it, reducing the amount of air that is reheated, recooled or mixed in each zone. This single-zone ventilation design in a multi-zone HVAC system offers significant opportunities for energy savings in any complex building. [Ref. C403.6.1]

REFRIGERATION SYSTEMS

The 2018 IECC continues the regulation of refrigeration equipment energy use. Tables C403.10.1(1) and (2) are organized by equipment type, operating mode and rating temperature and specify the maximum power use in kWh/day for the equipment. ASHRAE/IES Standard 90.1, which is adopted by reference as an alternative to the IECC—Commercial Provisions, also addresses the energy efficiency opportunities available from commercial refrigeration and freezing equipment. [Ref. C403.10] Walk-in coolers (Figure 8-12), walk-in freezers, refrigerated warehouse coolers and refrigerated warehouse freezers are regulated by these sections. Requirements include those related to doors; minimum floor, wall and ceiling insulation; anti-sweat heaters and controls and lighting efficiency.

Doors that do not exceed 45 inches wide or 7 feet high require automatic door closers that fully close doors to within 1 inch of full closure. Furthermore, all doorways are required to have strip doors, curtains, spring-hinged doors or another method of minimizing infiltration when doors are open. An example of a thermal consideration is the requirement for freezer floor insulation of R-28. Site-assembled walk-in coolers and freezers have similar requirements in the following subsection. **[Ref. C403.10.1]**

Site-assembled or site-constructed refrigerated display cases (Figure 8-13) require time switch or motion sensor controls for lighting. Time switch controls must turn off lights during nonbusiness hours, whereas motion sensor controls on each display case section must reduce lighting power by at least 50 percent within 3 minutes after the area within the sensor range is vacated. Low-temperature display cases must also incorporate temperature-based defrost termination controls with a time-limit default, and anti-sweat heater controls must be programmed to operate as a function of relative humidity. **[Ref. C403.10.2]**

You Should Know

Electrical power is generally rated in watts or horsepower. A horsepower is a unit of power equal to 746 watts. Most household hairdryers use about 1,400 watts to heat and blow warm air to dry your hair. Any VAV fan that uses about as much power as four household hairdryers is required to meet the efficiency requirements of Section C403.8.4.●

You Should Know

A fan efficiency grade (FEG) is a numerical rating that classifies fans by their aerodynamic ability to convert mechanical power to air power. The FEG allows for a quick comparison of fan efficiencies: more efficient fan models have higher FEG ratings. FEGs apply to the efficiency of the fan blades only and not to the motor and drives.●

FIGURE 8-12 Walk-in cooler. *(Photo courtesy of A S Klein Engineering, PLLC)*

Code Essentials

Definitions added to the 2018 IECC for mechanical systems

ISOLATION DEVICES: Devices that isolate HVAC zones so that they can be operated independently of one another. *Isolation devices* include separate systems, isolation dampers, and controls providing shutoff at terminal boxes.

NETWORKED GUESTROOM CONTROL SYSTEM: A control system, accessible from the front desk or other central location associated with a *Group R-1* building, that is capable of identifying the occupancy status of each guestroom according to a timed schedule, and is capable of controlling HVAC in each hotel and motel guestroom separately. ●

FIGURE 8-13 Refrigerated display case.

CHAPTER 9

Efficient Water Heating

Commercial building occupants have many uses for hot water. "Service water heating," as defined in the *International Energy Conservation Code®* (IECC®), is the "Supply of hot water for purposes other than comfort heating" **[Ref. C202]** and includes laundries, kitchens, swimming pools, showers and restrooms. The provisions of this section apply to commercial car washes, recreation centers and hot water produced in a building system used for any purpose other than comfort heating—the heating in HVAC, including car washes, recreation centers and any other defined commercial building.

A building's use is a very good indicator of service water-heating demand. Table 9-1 is published by the National Renewable Energy Laboratory (NREL) and illustrates estimated hot water consumption by building type.

The building service, pool and spa water-heating systems are required to be commissioned. The controls for the systems are of particular importance and integral to the efficient use of energy required to provide hot water for the systems. [Ref. C408]

TABLE 9-1 Water use table

Building type	Estimated hot water use Gallon/person/day
House	15.8
Hotel/Motel	20.0
Hospital	52.0
Office	1.1
Restaurant	2.4
School	0.5
School with showers	1.9

Courtesy: U.S. Department of Energy

EFFICIENCY

Table C404.2 of the IECC (see Table 9-2) lists minimum required efficiencies for water-heating equipment. Note that smaller-sized equipment in the same equipment type classification is tested to the DOE 10 CFR Part 430 standard. For more information on the standard, see the DOE section in Chapter 6 of the IECC. This is a reference to the National Appliance Energy Conservation Act (NAECA), an act of Congress that requires this federal standard to be satisfied for all water-heating equipment before it can be sold in the United States. Equipment efficiency must be indicated in the mechanical equipment schedule and comply with the performance minimums established in this table. [Ref. C404.2]

You Should Know

The National Appliance Energy Conservation Act (NAECA) of 1987 got its start 12 years earlier as the Energy Policy and Conservation Act (EPCA) in 1975. The EPCA established energy efficiency standards for major household appliances and required the U.S. Department of Energy (DOE) to begin the process of setting efficiency targets. This established national standards for many household appliances. The NAECA continues to evolve, with updates in the Energy Policy Acts of 1992 and 2005 and the Energy Independence and Security Act of 2007.

The NAECA now establishes minimum energy efficiency performance for residential and commercial appliances and consumer products, including office equipment and lighting products. The IECC references the minimum efficiencies for equipment regulated by the NAECA because it is a federal law; allowing anything less is a crime. However, exceeding the NAECA efficiencies is the basis for Option 1 in Section C406.1, which lists the additional efficiency package options. ●

TABLE 9-2 Minimum performance of water-heating equipment

Equipment type	Size category (input)	Subcategory or rating condition	Performance required[a,b]	Test procedure
Water heaters, electric	≤ 12 kW	Resistance	0.97 2 0.0003 V, EF	DOE 10 CFR Part 430
	> 12 kW	Resistance	$(0.3 + 27/V_m)$, %/h	ANSI Z21.10.3
	≤ 24 amps and ≤ 250 volts	Heat pump	2.057 - 0.00113V, EF	DOE 10 CFR Part 430
Storage water heaters, gas	≤ 75,000 Btu/h	≥ 20 gal	0.675 - 0.0015V, EF	DOE 10 CFR Part 430
	> 75,000 Btu/h and ≤ 155,000 Btu/h	< 4,000 Btu/h/gal	80% E_t (Q/800 1 110 $\sqrt{V}$) SL, Btu/h	ANSI Z21.10.3
	> 155,000 Btu/h	< 4,000 Btu/h/gal	80% E_t (Q/800 1 110 $\sqrt{V}$) SL, Btu/h	
Instantaneous water heaters, gas	50,000 Btu/h and < 200,000 Btu/h[c]	≥ 4,000 (Btu/h)/gal and < 2 gal	0.68 - 0.0019 V, EF	DOE 10 CFR Part 430
	≥ 200,000 Btu/h	≥ 4,000 (Btu/h)/gal and . 10 gal	80% E_t	ANSI Z21.10.3
	≥ 200,000 Btu/h	≥ 4,000 Btu/h/gal and ≥ 10 gal	80% E_t (Q/800 1 110 $\sqrt{V}$) SL, Btu/h	
Storage water heaters, oil	≤ 105,000 Btu/h	≥ 20 gal	0.59 - 0.0019 V, EF	DOE 10 CFR Part 430
	≥ 105,000 Btu/h	< 4,000 Btu/h/gal	80% E_t (Q/800 1 110. $\sqrt{V}$) SL, Btu/h	ANSI Z21.10.3

For SI: °C 5 [(°F) - 32J/1.8, 1 British thermal unit per hour 5 0.2931 W, 1 gallon 5 3.785 L, 1 British thermal unit per hour per gallon 5 0.078 W7L.

a. Energy factor (EF) and thermal efficiency (Et) are rninimum requirements. In the EF equation, V is the rated volume in gallons.

b. Standby loss (SL) is the maximum Btu/h based on a nominal 70°F temperature difference between stored water and ambient requirements. In the SL equation, Q is the nameplate input rate in Btu/h. In the SL equation for electric water heaters, V is the rated volume in gallons. In the SL equation for oil and gas water heaters and boilers, V is the rated volume in gallons.

c. Instantaneous water heaters with input rates below 200,000 Btu/h must comply with these requirements if the water heater is designed to heat water to temperatures 180°F or higher.

EFFICIENT HEATED WATER SUPPLY PIPING

Heated water supply piping limits the volume of piping between sources of heated water and the termination of fixture supply piping. Water heaters, circulating water systems and heat-trace temperature maintenance systems are considered sources of heated water. There are two methods of compliance. **[Ref. C404.5]**

The first is the Maximum Allowable Pipe Length method. The maximum allowable piping length from the nearest source of heated water to the termination of the fixture supply pipe must be in accordance with the maximum allowable length in Table C404.5.1, based on the nominal pipe size. There are two columns: one for public lavatory faucets and one for all other plumbing fixtures and appliances (Table 9-3).

TABLE 9-3 Table C404.5.1: Piping volume and maximum piping lengths

Nominal pipe size (inches)	Volume (liquid ounces per foot length)	Maximum piping length (feet)	
		Public lavatory faucets	Other fixtures and appliances
1/4	0.33	6	50
5/16	0.5	4	50
3/8	0.75	3	50
1/2	1.5	2	43
5/8	2	1	32
3/4	3	0.5	21
7/8	4	0.5	16
1	5	0.5	13
1 1/4	8	0.5	8
1 1/2	11	0.5	6
2 or larger	18	0.5	4

For SI: 1 inch = 25.4 mm, 1 foot = 304.8 mm, 1 liquid ounce = 0.030 L, 1 gallon = 128 ounces.

The second method of compliance is the Maximum Allowable Pipe Volume method. In this method, the volume from the nearest source of heated water to the termination of the fixture supply pipe is limited to not more than 2 ounces for a public lavatory faucet and ½ gallon for other plumbing fixtures or appliances.

Specific control functions are required on heated water circulating and temperature maintenance systems to limit energy losses. Details of each can be found in the subsections to Sections C404.6 and C404.7.

HEAT TRAPS

Hot water tends to rise in a storage vessel and can initiate a thermosyphoning loop as cool water replaces the moving warm water. Most commercial water-heating equipment has factory-installed heat traps. It is a good idea to specify this if you have the chance. Heat traps are required on hot water storage tanks but not on circulating service water-heating systems. **[Ref. C404.3]** Circulating hot water systems require a pump and loop back from the most distant fixture to the water heater. When the faucet is opened, hot water is immediately available. The pump can be activated by the push of a button, a timer or a motion-sensor switch. Controls are also required to limit the temperature of the water entering the cold-water piping to no more than 104°F. **[Ref. C404.7]**

PIPE INSULATION

Just like piping serving as part of a heating or cooling system, piping from water heaters to the termination of the heated water fixture supply pipes must be insulated in accordance with Table C403.11.3 in the mechanical section of the code.

Water-heating piping in circulating systems must be insulated because the pipe is full of hot water most of the time. Piping serving water heaters that do not have an integral heat trap must have insulation installed on the first 8 feet of pipe leading in and out of the water heater or to a heat trap, whichever is less (Figure 9-1).

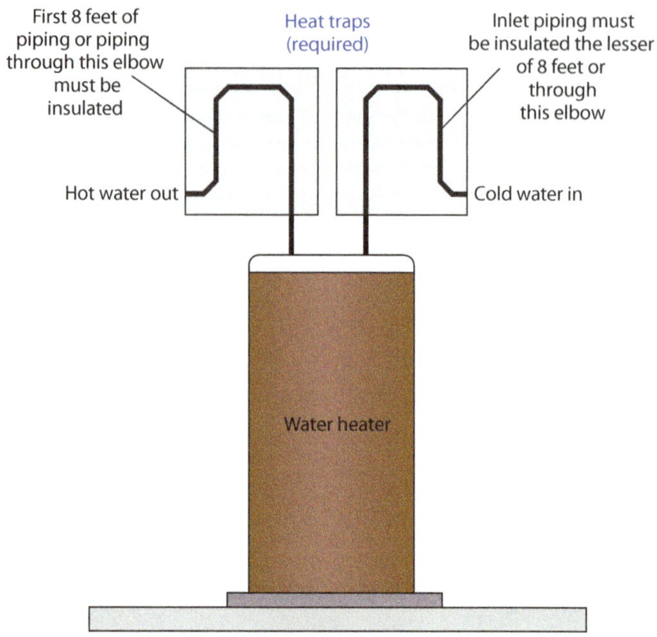

FIGURE 9-1 Insulation requirements for water heaters.

Heat-trace pipe systems have an electric heat tape installed in contact with the water pipe. The heat trace maintains the desired water temperature in the pipe and is an alternative to circulating systems. Insulation requirements are specified by the manufacturer or must be in accordance with Table C403.11.3. **[Ref. C404.4]**

Both circulating and heat-trace hot water piping systems are required to have a control that will turn off the system when hot water demand is low. **[Ref. C404.6]** Service water-heating systems are required to undergo commissioning in accordance with Section C408.2.

POOL AND SPA HEATERS AND COVERS

Swimming pool and spa operations use energy to heat and filter water. Heat energy is lost when the air temperature at the water surface is lower than the pool or spa water temperature and through evaporation (Figure 9-2). Specific energy-saving provisions apply to pool and spa heaters, time switches and covers.

FIGURE 9-2 Spa cover.

You Should Know

Energy is required to change the state of water from a liquid to a vapor. The amount of energy it takes for 1 ounce of water to evaporate from a pool or spa decreases the temperature of about 2.5 gallons of the remaining water by 1°F. An everyday example of this phenomenon is you standing in front of a fan after showering. You feel cold because the evaporation of water from your skin requires accepting heat energy from your body. ●

Heaters must have an on/off switch mounted outside of the heater that can be reached easily. This allows the heater to be turned off when the pool may not be used for days at a time. Time switches that can automatically turn the heater and pumps on and off according to a preset schedule are required for all installations, with exceptions for public health standards and solar- and waste-heat recovery pumps.

Water evaporates. Pool and spa owners should expect ¼ to ⅜ inch of water loss daily from evaporation, and this is sometimes mistaken for a leak. Evaporation can be significantly reduced with the vapor-retardant pool cover required by the IECC (Figure 9-3). A cover is not required for pools and spas getting more than 75 percent of seasonal heating energy from site-recovered energy such as a heat pump or thermal solar system. **[Ref. C404.9.3]**

FIGURE 9-3 Pool cover

The energy consumption of electric-powered portable spas must meet the provisions of APSP 14. This is a mandatory requirement. **[Ref. C404.10]**

You Should Know

The 2012 IECC provisions only applied to pools and permanently installed in-ground spas. A standard is published that addresses portable electric spas. Jurisdictions adopting the 2012 *International Swimming Pool and Spa Code®* (ISPSC®) find reference to the APSP 14-2011, *American National Standard for Portable Electric Spa Energy Efficiency*, in Section 303, titled "Energy." The energy provisions in the ISPSC are very similar to those in the IECC, with the exception of portable residential spas. Those units must comply with APSP-14 in jurisdictions administering the 2012 ISPSC. The 2015 IECC began the governing of portable electric spas by its own reference to APSP 14, which continues in the 2018 edition of the IECC.

The requirements for pool and spa covers are worded differently in the IECC and ISPSC, but accomplish the same intent, which is to prevent evaporation. ●

CHAPTER 10

Electric Power and Light

Outside of the heating, cooling and ventilation, lighting contributes a significant amount towards the energy consumption of a commercial building. While lighting fixtures have seen many advances in technology, lighting controls offer opportunities for saving energy without the need to change occupant behavior, so in a way the reductions are automatic and mostly unnoticed. Efficient lighting sources and controls maintain and may even improve interior and exterior illumination. Numerous studies suggest that daylighting improves employees' satisfaction with working conditions and reduces absenteeism. Careful attention to lighting design and control benefits building occupants and building owners, therefore, most of the provisions in this section of the IECC are mandatory.

You Should Know

High-efficacy lamps are defined as compact fluorescent lamps, T-8 or smaller-diameter linear fluorescent lamps, or lamps with a minimum efficiency of:

1. 60 lumens per watt for lamps over 40 watts;
2. 50 lumens per watt for lamps over 15 watts to 40 watts; and
3. 40 lumens per watt for lamps 15 watts or less.

Figure 10-1 offers efficacy comparisons using this definition and Edison's lightbulb. ●

APPLICABILITY

The provisions in this code section apply to lighting systems in new buildings, additions, tenant finishes, alterations to existing lighting systems and any change in occupancy. Controls and minimum lamp efficiencies are required for interior and exterior lighting systems. Exceptions to the provisions include some historic buildings, replacing less than 10 percent of the existing interior luminaires and alterations that do not increase the required lighting power. Lighting in dwelling units in a commercial building is also exempt provided 90 percent of the installed fixtures and lamps are high-efficacy devices. **[Ref. C405.1]**

LIGHTING CONTROLS

Interior lighting systems must be designed with the intent that lights in the space are on when they need to be and only as many lights are on in the space as needed to make it safe and functional. In areas of a building under skylights and spaces near walls with lots of windows, lighting may not be needed during much of the day. In fact, in these conditions it may be difficult to know if the lights are on at all. In some cases, these daylight zones must have controls capable of dimming and brightening the space separate from other interior lighting. Exterior lighting control provisions allow for times when the building is used after dusk and before dawn, again to make the areas safe and functional. **[Ref. C405.2]**

Interior

Interior controls are required in every space enclosed by floor-to-ceiling partitions and the 2018 edition of the IECC brings in control requirements for open office spaces that utilize large open spaces instead of individual offices, requiring that the space be split up into individual lighting zones no greater than 600 square feet in floor area to more efficiently control the lighting uses. **[Ref. C405.2.1.3]** In some spaces, the control may be as simple as a user-friendly on/off light switch in the space. An occupancy sensor or time switch to activate and deactivate the lights in the room may also comply. In warehouses, the lighting in aisleways and open areas must be controlled with occupant sensors that reduce the lighting power by 50 percent when the areas are not occupied. **[Ref. C405.2.1.2]**

There are three important and logical exceptions to the requirements concerning individual space and occupant-accessible controls: Emergency and security areas that must be continuously lighted, stairways and corridors that are part of the means of egress do not require lighting controls and emergency egress lighting that is normally off.

LAMP	EFFICACY LUMENS/WATT
Edison's first lamp	1.4
Incandescent	10–40
Halogen	20–45
Fluorescent	35–100
Mercury	50–60
Metal halide	80–115
High pressure sodium	100–140

(Courtesy of U.S. Department of Energy, Office of Building Technology, State and Community Programs, www.energycodes.gov)

FIGURE 10-1 Efficacy comparison.

An LLLC has the following **three equipment components:**

1. Single multi-type sensor (occupancy and photocell)
2. Luminaire controller
3. Continuous dimming ballast/driver

} All three located at/in each luminaire

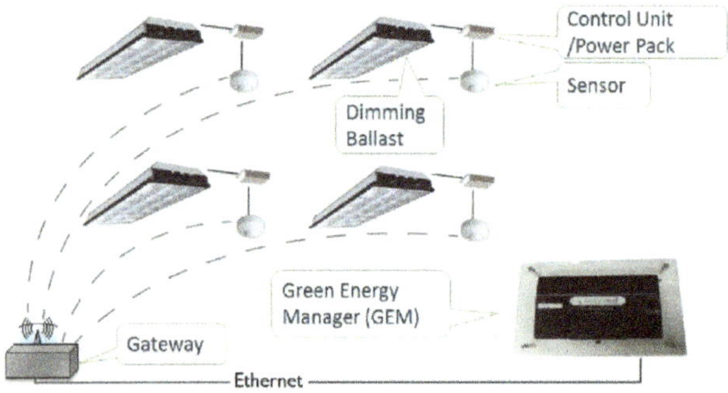

Plus the use of:

4. Gateway to transfer information
5. Software package which provides simple commissioning and user interface
6. Monitoring and data collection software for occupancy and lighting energy consumption, with sampling typically once per minute

FIGURE 10-2 Luminaire-Level Lighting Controls

Reducing the amount of lights that are on in a space can save quite a bit of energy, thus the requirements for light reduction controls. The idea is not to have one-half of the room bright and the other half dark, but to evenly reduce the light in the entire room. The IECC requires any space that is required to have light reduction controls to have a manual control that allows the occupant to reduce the lighting load by at least 50 percent and gives several options for doing just that (Figure 10-2). The lighting design and fixture schedule will confirm compliance with this provision.

Reduction controls

Spaces required to have manual controls must also have a strategy to reduce the lighting load by 50 percent. The idea is not to have one-half of the room bright and the other half dark, but to evenly reduce the light in the entire room. The lighting design and fixture schedule will confirm compliance with this provision. Figure 10-2 illustrates three common methods of compliance. One way is to dim all the lights. This does not work with all types of luminaires. Another approach is to wire two circuits, with each switch controlling half of the light fixtures. The final two-switch option allows one switch to turn on half of the lamps in each fixture and the second switch to turn on the other half of the lamps in the fixtures.

The IECC notes specific areas and designs that are exempt from the manual lighting reduction control provisions. These exceptions are

You Should Know

Luminaire-level lighting controls, as defined in C202, is "A lighting system consisting of one or more luminaires with embedded lighting control logic, occupancy and ambient light sensors, wireless networking capabilities and local override switching capability, where required."

In the 2018 IECC, these controls are presented as an option to occupant sensor controls and time switches. The controls monitor ambient light, both electric light and daylight, and brighten or dim the artificial light to maintain a desired light level. They are also capable of monitoring occupant activity to brighten a space when it is occupied or to dim it when unoccupied. **[Ref C05.2]** ●

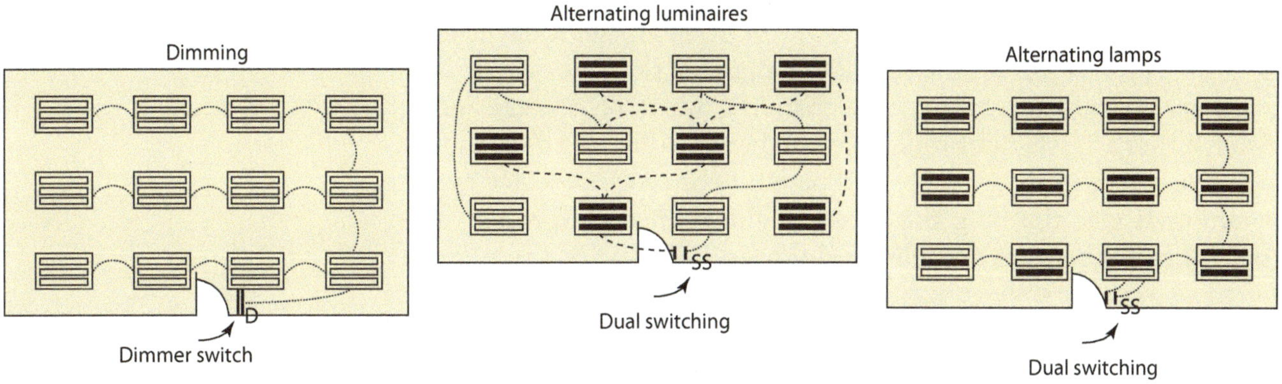

FIGURE 10-3 Typical open office space.

> **Code Essentials**
>
> "Ready Access (to)": That which enables a device, appliance or equipment to be directly reached, without requiring the removal or movement of any panel or similar obstruction. ●

areas that have only one luminaire with less than a 100-watt power rating, corridors, building equipment and service rooms, sleeping units, spaces that use less than 0.6 watts per square foot and compliant daylight spaces. **[Ref. C405.2.2.1, Exception 2]**

Additional control specifics

Every area in a commercial building project regulated by the provisions of this code is also required to have one or more of three types of additional controls to reduce lighting when spaces are not in use or when there is sufficient natural lighting. Exempted areas are sleeping units, rooms lighted for patient care, areas that could be dangerous when the lights are off and rooms lighted for continuous use. Automatic time-switch controls satisfy this requirement because they turn off the lighting system when the building space is not in use. The control must be installed so that it has *ready access (to)* (see the definition **[Ref. C202]**) and the lights it controls must be viewable from the switch. Malls, arcades, auditoriums, arenas, some very large buildings, emergency egress lighting and spaces with occupancy sensors or luminaire-light level controls are exempt from these specific provisions.

Occupancy sensors are simple and effectively reduce energy use. These controls turn off the lights after a set time if occupants forget to do so as they leave the room. Sensors are logically required in warehouses, classrooms, conference and meeting rooms, office break rooms, private offices, open offices, storage rooms, janitorial closets, restrooms, locker rooms and spaces less than 300 square feet in area with floor-to-ceiling walls. These controls must turn the lights off within 20 minutes of the people leaving. **[Ref. C405.2.1]**

Daylight zones

Daylighting design strategy makes the best use of available sunlight by thoughtful and calculated window and skylight placement. These areas are called daylight zones and, when properly designed, require minimum artificial light. Lighting in daylight zones must be controlled independently of the general lighting system. This allows functional light on bright, sunny days as well as on dark and cloudy ones. The area of a daylight zone depends on the size and number of skylights and windows in or next to the space. A daylight zone can be on any level of the building, not just the areas on the top floor beneath a skylight.

In spaces with a total of more than 150 watts of general lighting within either a sidelight or toplight daylight zone (Figure 10-3), the controls must have ready access to and have a light-sensing device capable of automatically dimming the general lighting in each daylight zone. Daylight zones beneath skylights and those adjacent to windows are required to have separate controls. Furthermore, sidelight daylight zones are required to have separate controls for zones facing different cardinal directions (i.e., within 45 degrees of due north, south, east or west). Exceptions are given to sidelight daylight zones on the first floor above grade in Group A-2 and Group M occupancies, allowing restaurants and shops in strip centers, for example, to be exempt. **[Ref. C405.2.3]**

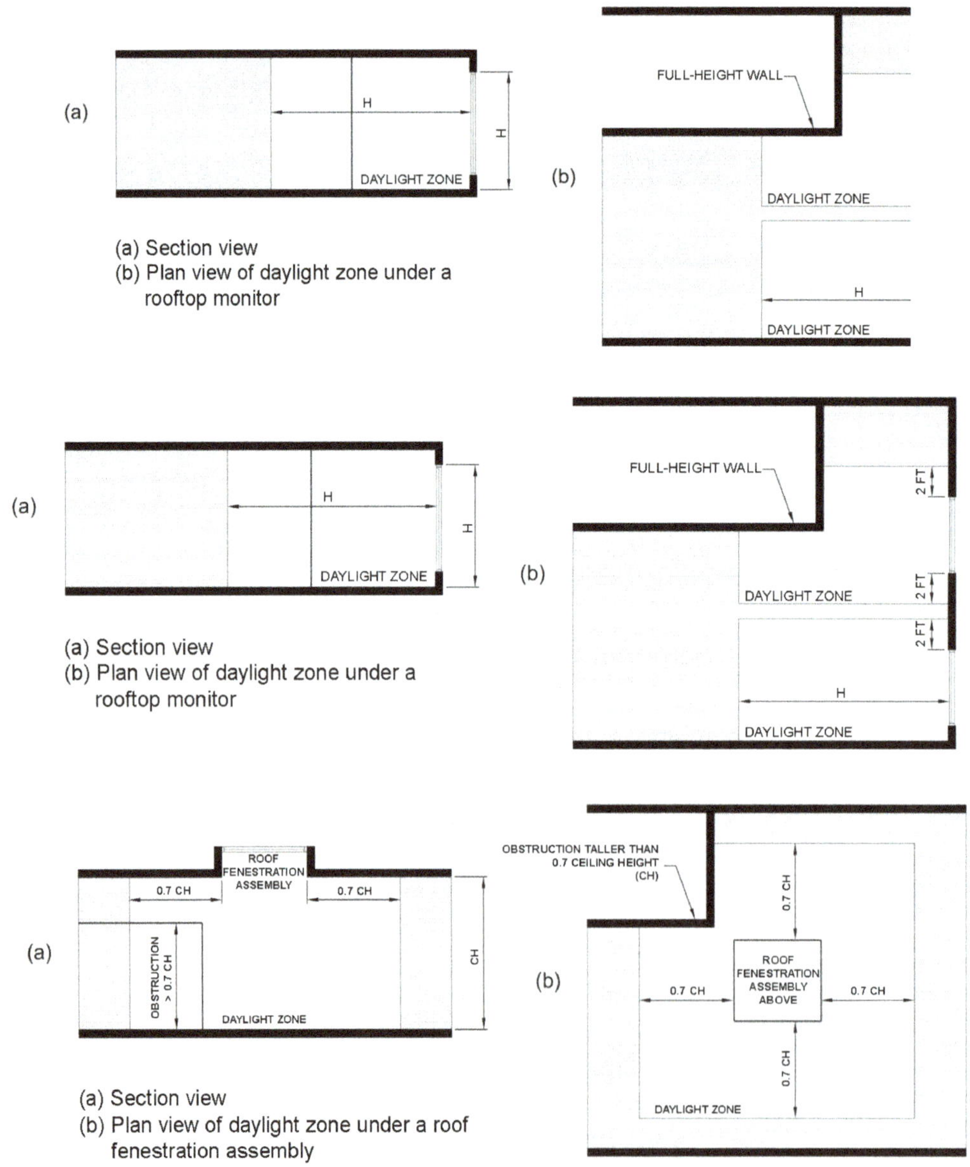

CH = ceiling height
H = head height

FIGURE 10-4 Daylight zones under a skylight, under a rooftop monitor and next to a window.

You Should Know

"Lighting power density" (LPD) is not a defined term in the IECC but is a simple concept best understood by its mathematical expression:

LPD = watts of lighting power divided by the square foot area of the space considered, or w/ft^2.

The lighting power of each luminaire is determined in accordance with Equation 4-10 in Section C405.3.1 based on the type of lighting.

This information comes directly from the lighting schedule. The square foot number is the size of the building using the whole-building-area method or the sum of the square feet for the specific space most closely described by the type of activity and LPD number using the space-by-space method. ●

Specific application lighting controls

Additional controls are required in other specific described instances. Display and accent lighting, such as that used for pictures in hotel or office lobbies and lighting in display cases, must be controlled separately from the other lighting in the space. Hotel and motel rooms and suites must have a control device at the main entry door that operates the permanent light fixtures and switched receptacles in the room or suite. Supplemental task lighting over desks and work surfaces and under counters must have a switch on the fixture or mounted nearby and accessible on the wall. Artificial plant-growth lights and food-warming lights are considered lighting for nonvisual applications. These types of lighting must have a control separate from those of the other lighting in the same room. Lighting fixtures that are displayed for sale and lighting equipment used for education and demonstration also must be controlled separately from the other lighting in the space. **[Ref. C405.2.4]** Exterior building lighting must turn the lights off when daylight is present and satisfies the lighting needs. **[Ref. C405.2.6.1]**

Interior lighting power

The energy code offers two paths of compliance with the interior lighting requirements. Not all lighting equipment is included in the calculations, and most of the exceptions are similar to those in the controls section. The 19 exceptions are listed in Section C405.3.1. Professional indoor sports arenas and fields and casino gaming areas are examples of exceptions introduced here to the indoor lighting provisions, which makes perfect sense given their size.

The building area method limits lighting power density (LPD) by first describing the building area and associating the allowable watts per square foot shown in Table C405.3.2(1) in the IECC. This LPD times the building floor area is the interior lighting power allowance for each noncontiguous building area type. Exempted building area and lighting equipment in those areas is not included in the calculations, and no trade-offs are permitted between areas. Equation 4-10 in Section C405.3.1 is used to determine the value used to calculate the proposed lighting power. This information is best described in the lighting plan and lighting fixture schedule.

The space-by-space method refers to the building areas as best described by uses in Table C405.3.2(2) in the IECC (Table 10-1). Each area of use is associated with a value in the table. Each area of use times the LPD value of all the building spaces equals the interior power allowed for compliance. This method allows trade-offs between spaces to use more or less power to comply if the total building design is less than or equal to the total allowed in the whole building.

EXAMPLE:

Q: The lighting plan for a 1200-square-foot office with no floor-to-ceiling partitions proposes recessed-can light luminaires rated at 50 watts each and a 15-foot track light. The lighting designer specifies 10 30-watt lamps in each recessed luminaire and 10 40-watt fixtures in the track. What is the LPD for the office space?

A: The total labeled wattage for recessed luminaires is 50 × 10 luminaries = 500 watts. The total wattage of the track lights is 40 × 10 fixtures = 400 watts, but that is less than the minimum connected power of 30W/lin ft. (30W × 15ft = 450 W). The total connected power is 500W + 450W = 950W. When this is divided by 1,200 square feet, the LPD is 0.792 LPD.

Q: Does this space comply with the space-by-space method?

A: Yes. Table C405.3.2(2) allows up to 0.81 LPD for an open office, so the office space complies.

TABLE 10-1 Table C405.3.2(2): Interior lighting power allowances: space-by-space method

Common space types[a]	LPD (watts/sq.ft)
Atrium less than 40 feet in height	0.03 per foot in total height
Atrium greater than 40 feet in height	0.40 + 0.02 per foot in total height
Audience seating area	
In an auditorium	0.63
In a convention center	0.82
In a gymnasium	0.65
In a motion picture theater	1.14
In a penitentiary	0.28
In a performing arts theater	2.03
In a religious building	1.53
In a sports arena	0.43
Otherwise	0.43
Banking activity area	0.86
Breakroom (See Lounge/Breakroom)	
Classroom/lecture hall/training room	
In a penitentiary	1.34
Otherwise	0.96
Conference/meeting/multipurpose room	1.07
Copy/print room	0.56
Corridor	
In a facility for the visually impaired (and not used primarily by the staff)[b]	0.92
In a hospital	0.92
In a manufacturing facility	0.29
Otherwise	0.66

(Continues)

TABLE 10-1 Table C405.3.2(2): Interior lighting power allowances: space-by-space method *(Continued)*

Common space types[a]	LPD (watts/sq.ft)
Courtroom	1.39
Computer room	1.71
Dining area	
In a penitentiary	0.96
In a facility for the visually impaired (and not used primarily by the staff)b	2.00
In bar/lounge or leisure dining	0.93
In cafeteria or fast food dining	0.63
In family dining	0.71
Otherwise	0.63
Food preparation area	1.06
Guestroom	0.47
Laboratory	
In or as a classroom	1.43
Otherwise	1.81
Laundry/washing area	0.6
Loading dock, interior	0.77
Lobby	
In a facility for the visually impaired (and not used primarily by the staff)[b]	1.8
For an elevator	0.64
In a hotel	1.06
In a motion picture theater	0.59
In a performing arts theater	2.0
Otherwise	0.9
Locker room	0.75
Lounge/breakroom	
In a healthcare facility	0.92
Otherwise	0.73
Office	
Enclosed	1.11
Open plan	0.98
Parking area, interior	0.19
Pharmacy area	1.68
Restroom	
In a facility for the visually impaired (and not used primarily by the staffb	1.21
Otherwise	0.98
Sales area	1.59
Seating area, general	0.54
Stairway (See space containing stairway)	
Stairwell	0.69
Storage room	0.63
Vehicular maintenance area	0.67
Workshop	1.59
Building type specific space types[a]	**LPD (watts/sq.ft)**
Facility for the visually impaired[b]	
In a chapel (and not used primarily by the staff	2.21
In a recreation room (and not used primarily by the staff)	2.41

(Continues)

TABLE 10-1 Table C405.4.2(2): Interior lighting power allowances: space-by-space method (*Continued*)

Building type specific space types[a]	LPD (watts/sq.ft)
Automotive (See Vehicular maintenance area above)	
Convention Center—exhibit space	1.45
Dormitory—living quarters	0.38
Fire Station—sleeping quarters	0.22
Gymnasium/fitness center	
In an exercise area	0.72
In a playing area	1.2
Healthcare facility	
In an exam/treatment room	1.66
In an imaging room	1.51
In a medical supply room	0.74
In a nursery	0.88
In a nurse's station	0.71
In an operating room	2.48
In a patient room	0.62
In a physical therapy room	0.91
In a recovery room	1.15
Library	
In a reading area	1.06
In the stacks	1.71
Manufacturing facility	
In a detailed manufacturing area	1.29
In an equipment room	0.74
In an extra high bay area (greater than 50¢ floor-to-ceiling height)	1.05
In a high bay area (25-50¢ floor-to-ceiling height)	1.23
In a low bay area (less than 25′ floor-to-ceiling height)	1.19
Museum	
In a general exhibition area	1.05
In a restoration room	1.02
Performing arts theater—dressing room	0.61
Post Office—Sorting Area	0.94
Religious buildings	
In a fellowship hall	0.64
In a worship/pulpit/choir area	1.53
Retail facilities	
In a dressing/fitting room	0.71
In a mall concourse	1.1
Sports arena—playing area	
For a Class I facility	3.68
For a Class II facility	2.4
For a Class III facility	1.8
For a Class IV facility	1.2

(*Continues*)

TABLE 10-1 Table C405.4.2(2): Interior lighting power allowances: space-by-space method (*Continued*)

Building type specific space types[a]	LPD (watts/sq.ft)
Transportation facility	
In a baggage/carousel area	0.53
In an airport concourse	0.36
At a terminal ticket counter	0.8
Warehouse—storage area	
For medium to bulky, palletized items	0.58
For smaller, hand-carried items	0.95

a. In cases where both a common space type and a building area specific space type are listed, the building area specific space type shall apply

b. A 'Facility for the Visually Impaired' is a facility that is licensed or will be licensed by local or state authorities for senior long-term care, adult daycare, senior support or people with special visual needs.

The space-by-space method permits an increase in interior lighting power allowance for specific lighting functions in the sales area in retail space. The basic allowance of 1.59 watts per square foot for a sales area may be increased by the area allotted to specific classes of merchandise according to Equation 4-11 in Section C405.3.2.2.1. When these allowances are taken, the additional lighting circuits must be switched or dimmed separately from the general lighting. [Ref. C405.3.2.2.1]

EXTERIOR LIGHTING

FIGURE 10-5 Exempt exterior lighting.

Lighting installed in parking lots, outside building entrances, on the walkways connecting the parking areas and highlighting building façades is regulated by the provisions in this section. Lighted advertising and traffic directional signs are exempt, as are solar lights not connected to the building power source. Lighting for public monuments (Figure 10-4) and exterior features of historic landmark structures and buildings is also exempted.

The basic power allowance and LPD for specific exterior lighting applications is determined by Table C405.4.2(1) in the IECC, which describes lighting zones (Table 10-2). After choosing the most appropriate zone considering the building's location, calculate the total exterior lighting power according to Table C405.4.2(2). These lighting power allowances are like the space-by-space LPD calculation. Once the allowed total is established, the designer may use more light in one area and less in another, and still comply if the total is equal to or less than the budget allowed.

Individual lighting power allowances are allowances for additional lighting of surfaces like security lighting that needs to be on all night, such as at ATMs, ambulance entries and inspection stations at guarded entries. This extra allowance can only be used for the specified purposes and cannot be traded to other surfaces. **[Ref. C405.4.2.1 and Table C405.4.2(3)]**

TABLE 10-2 Table C405.4.2(1): Exterior lighting zones

Lighting zone	Description
1	Developed areas of national parks, state parks, forest land, and rural areas
2	Areas predominantly consisting of residential zoning, neighborhood business districts, light industrial with limited nighttime use and residential mixed use areas
3	All other areas
4	High-activity commercial districts in major metropolitan areas as designated by the local land use planning authority

ELECTRIC MOTORS AND TRANSFORMERS

Two mandatory sections state that electrical transformers and electric motors must meet the minimum efficiency requirements of Tables C405.6 and C405.7(1-4). Efficiencies must be tested and rated in accordance with the procedure in DOE 10 CFR 431, and ratings data are usually furnished by the motor manufacturer Table 10-3 (2018 IECC Table C405.6) shows efficiency levels of 10 CFR 431. Several exemptions for transformers exist, including special-purpose transformers and those that meet exclusions in federal regulations. **[Ref. C405.6-7]**

TABLE 10-3 Minimum efficiency tables for transformers and motors

Lighting zone	Description
Table C405.7	Low-voltage dry-type distribution transformers
Table C405.8(1)	60 Hz NEMA general purpose (subtype I) motors
Table C405.8(2)	General purpose (subtype II) and Design B > 200 hp motors
Table C405.8(3)	Polyphase motors
Table C405.8(4)	Capacitor-start capacitor-run and induction-run motors

ELEVATORS, ESCALATORS AND MOVING WALKS

Elevators, escalators and moving walks are regulated through the IECC Commercial Electrical Power and Lighting Systems provisions. Luminaires in each elevator cab shall be not less than 35 lumens per watt, and ventilation fans in elevators that do not have their own air-conditioning system must not consume more than 0.33 Watts per cubic foot per minute. Controls must be provided that will de-energize ventilation fans and lighting systems when the elevator is stopped, unoccupied and with its doors closed for over 15 minutes. **[Ref. C405.8.1]**

Escalators and moving walks are required to be equipped with automatic controls configured to reduce speed to a minimum when not conveying passengers. Reversible escalators or those designed for down operation must have a variable frequency regenerative drive that supplies electrical energy to the building electrical system when the escalator is loaded with passengers whose combined weight exceeds 750 pounds. Figures 10-6 and 10-7 show examples of commercial escalators that must comply with IECC energy efficiency provisions. **[Ref. C405.8.2]**

FIGURE 10-6 Double escalators at an airport facility.

FIGURE 10-7 Escalator capable of reducing speed.

CHAPTER

11 Required Efficiency Options

ADDITIONAL EFFICIENCY FEATURES

The additional efficiency provision of Section C406 in the *International Energy Conservation Code*® (IECC®) completes the prescriptive path of energy code compliance for commercial buildings. Eight options complement the minimum mandatory provisions for building envelope performance, mechanical systems, service water heating and lighting efficiency. **[Ref. C406.1]** ASHRAE 90.1-2016 and the performance path (Section C401.2, Items 1 and 3) building designs produce similar overall results in effective and efficient energy use and, therefore, do not have to comply with Section C406.

It is important to understand that these additional efficiency package options are not options that take the building above the minimum requirements of the energy code, but instead help the building to meet the minimum requirements. Instead of going through the envelope, mechanical, service water heating and lighting requirements of the IECC prescriptive approach and tightening everything up by 10-15 percent, the decision was made to give the building owner and designer the ability to decide how best to make their building or system more efficient, thus keeping with the IECC's original intent of providing flexibility in order to conserve energy in buildings. **[Ref. C101.3]**

HVAC system performance

The first option results in more efficient heating, ventilation, and air-conditioning (HVAC) performance than required by Section C403. If this option is chosen, the provisions for mechanical systems still apply, and the HVAC equipment efficiency must exceed that listed in Tables C403.3.2 (1–7) by 10 percent. Variable refrigerant flow systems are required to exceed the energy efficiency provisions of ASHRAE 90.1 by 10 percent. Equipment not listed in Table C403.3.2(7) is limited to 10 percent of the total building system capacity. [Ref. C406.2]

Lighting power density

The second option reduces the lighting power density (LPD) to 90 percent of that allowed by either the Building Area Method or the Space-by-Space Method of Section C405.3.2. Compliance requires that the building space functions are identified and measured, and the supporting LPD calculations must be submitted with the plans to demonstrate compliance. [Ref. C406.3]

Lighting controls

The third option is enhanced digital lighting controls. Luminaires are required to be capable of continuous dimming and being addressed individually. Fixtures are required to be controlled through a digital control system. Construction documents must include the sequence of operations for the control system, and functional testing must be performed during commissioning. [Ref. C406.4]

On-site renewables

Two renewable system choices are available to comply with the fourth option of this provision. A photovoltaic (PV) system (Figure 11-1) providing at least 0.50 watts or a solar thermal (hot water) system providing at least 1.71 Btu per square foot of the building conditioned floor area satisfies this option. Option 2 requires that at least 3 percent of the energy used by the mechanical, service hot water and lighting systems in the building be provided by on-site renewables. This allows flexible design to maximize the local solar resources to fit the building's energy demands. [Ref. C406.5]

FIGURE 11-1 Photovoltaic (PV) and solar thermal (hot water) systems

You Should Know

"Variable refrigerant flow system" is defined in Section C202 as "An engineered direct-expansion (DX) refrigerant system that incorporates a common condensing unit, at least one variable-capacity compressor, a distributed refrigerant piping network to multiple indoor fan heating and cooling units each capable of individual zone temperature control, through integral zone temperature control devices and a common communications network. Variable refrigerant flow utilizes three or more steps of control on common interconnecting piping.●

Dedicated outdoor air system

The fifth option is a dedicated outdoor air system (DOAS) that is available to buildings with hydronic and multiple-zone HVAC systems. If this option is chosen, such systems shall be equipped with an independent ventilation system designed to provide not less than the minimum 100-percent outdoor air to each individual occupied space, as specified by the *International Mechanical Code®* (IMC®). Furthermore, the HVAC system must include automatic controls that are configured to reset the supply-air temperature in response to building loads or outdoor air temperatures. **[Ref. C406.6]**

You Should Know

Dedicated outdoor air system (DOAS) is not a defined term in the IECC but is best described as a dedicated system to handle the building's ventilation needs rather than having ventilation be a part of the building's space conditioning system. A DOAS handles the outdoor air latent load in addition to some of the sensible load, and a parallel system handles the rest of the sensible load. ●

Service water heating

The sixth option is reduced energy use in service water heating. Buildings shall be of certain types to use this compliance method (Table 11-1) or show a service hot water load of 10 percent or more of total building energy loads, as indicated with an energy analysis. If a building meets the requisites and this option is chosen, then not less than 60 percent of hot water requirements must be supplied by waste heat recovery or solar water-heating systems. **[Ref. C406.7]**

Enhanced envelope performance

The seventh option, new to the 2018 edition of the code, is to have the total UA of the building thermal envelope shown to be 15 percent below the total UA of the envelope if it were done in accordance with Section C402.1.5. **[Ref. C406.8]**

Reduced air infiltration

Also new to the 2018 IECC is the eighth and final option of proving that the building is air tight by performing a whole-building pressurization test in accordance with either ASTM E779 or ASTM E1827. If choosing this option, the air-leakage rate of the envelope cannot exceed 0.25 cfm/ft^2 at 75 Pascals. The surface area must include the above- and below-grade building envelope. An exception exists for testing the whole building once the conditioned floor area exceeds 250,000 square feet, allowing for testing to be conducted on just above-grade sections of the building. **[Ref. C406.9]**

TABLE 11-1 Specific building use types allowed to use the additional compliance option in Section C406.7

1	Group R-1	Boarding houses, hotels or motels
2	Group I-2	Hospitals, psychiatric hospitals and nursing homes
3	Group A-2	Restaurants and banquet halls or buildings containing food preparation areas
4	Group F	Laundries
5	Group R-2	Buildings with residential occupancies
6	Group A-3	Health clubs and spas

PART

IV

General Residential Energy Provisions

CHAPTER

12 General Provisions

The provisions of Sections R101 and R102 of the *International Residential Code*® (IRC®) establish a framework to address the administration, application and enforcement of the residential energy code provisions for one- and two-family dwellings and townhouses. These sections address requirements for organization of the building plans and information needed in the supporting documents. Many states require the construction documents to be prepared by a registered design professional, usually an architect or engineer. Homeowners may be allowed to submit building plans and build their own homes. It is advisable to check with the code official in the jurisdiction to get specific information about this process.

The building contractor's responsibilities regarding permits and inspections and the code official's responsibilities concerning plan review and inspections are individually addressed and detailed in subsequent sections in Part II of the IRC. These sections describe the relationships and understandings between the building department authority and the design and construction community.

SCOPE

The residential provisions apply only to residential buildings as defined in the IECC. This distinction is not as obvious as it seems. Entire buildings and building sites or just parts of building systems and building equipment may be classified as residential. **[Ref. R101.2]** It is essential to know the difference in residential and commercial classifications as applied in the IECC (Figure 12-1). As stated in Section R202, the definition of "*Residential building*" is as follows: "For this code, includes detached one- and two-family dwellings and townhouses as well as *Group* R-2, R-3 and R-4 buildings three stories or less in height above grade plane," as shown in Figure 12-2." "*Grade plane*" is defined in Section 202 of the *International Building Code*® (IBC®). As noted earlier in this text, an italicized term in the code means that the word or phrase has a specific meaning in code language, and a definition is provided to clarify its intended use. Chapter 2[RE] of the IECC lists 70 words and specific definitions to establish the common vocabulary for the residential energy provisions.

Code Essentials

The residential provisions of the IECC are duplicated, verbatim, in the *International Residential Code* (IRC). The IRC is intended to be a standalone document that contains all provisions applicable to buildings within its scope: detached one- and two-family dwellings and multiple single-family dwellings (townhouses) not more than three stories above grade plane in height with a separate means of egress. The residential portion of the IECC applies to a broader set of buildings than the IRC.

FIGURE 12-1 Three-story multi-family residential units are regulated by the residential provisions in the IECC.

Classifications and descriptions of residential occupancies are found in IBC Section 310. Apartments, condominiums and timeshare buildings are typical R-2 occupancies. Single-family homes, duplexes and townhomes are R-3 buildings. The shared attribute of these residential uses is that the occupants are "nontransient." Thus, if the building uses fit into any of the defined descriptions of "Residential" and the building is less than four stories in height above grade, the IECC—Residential provisions apply. **[Ref. R101.2]**

INTENT

As noted in Chapter 3 of this text, the intent of the IECC is stated simply: "This code shall regulate the design and construction of buildings for the effective use and conservation of energy over the useful life of each building." *Webster's Dictionary* defines *shall* as "used in laws, regulations, or directives to express what is mandatory." All the provisions and regulations must pass this test of intent to conserve energy use through the design and construction of buildings to be included in the code. The design requirements of the energy code are intended to promote the effective use of energy in new buildings and building systems as well as in repairs and other work undertaken in existing buildings, "over the useful life of each building." The term "useful life" is a somewhat ambiguous term, as many factors may affect a building's life span. In the case of residential buildings, many are only meant for a useful life of 30 years or so, the length of a typical mortgage, whereas others have existed for a century or more and may be classified as historic buildings.

FIGURE 12-2 Single-family residences are regulated by the provisions in the IECC.

Also, as noted in the commercial provisions discussion (Chapter 3), the intent of the code is to encourage—not stifle—the use of new and innovative energy-saving devices and designs, and, therefore, materials and techniques not specifically addressed in the IECC may be approved by the code official if they contribute to the overall intent of conserving energy. [Ref. R101.1]

You Should Know

"Owner-occupied lodging houses" is a residential category that was added in the 2012 IRC. These are commonly known as bed-and-breakfasts and can be constructed under the provisions of the IRC if they have five or fewer guestrooms and are equipped with a fire sprinkler system in accordance with IRC Section P2904.

APPLICABILITY

As in the commercial provisions, the residential regulations of the IECC note both general and specific requirements for certain conditions. In general, when two different code provisions apply to the same condition, the most restrictive requirement applies. However, in situations where the difference is between a general requirement and a specific requirement, the specific provision applies to the design and construction, as is true of the other International Codes® (I-Codes®). [Ref. R101.4]

LOW-ENERGY BUILDINGS

Low-energy residential buildings are exempt from the thermal envelope code requirements. The building or portion of the building must be separated from the conditioned areas with walls or ceilings that comply with the thermal envelope provisions. For example, an attached garage with a space-heating unit intended to keep exposed pipes from freezing may qualify for this exemption. The maximum heating capacity allowed is 1.0 watt/ft^2 or 3.4 Btu/h • ft^2. Low-energy buildings are not exempt from the mandatory lighting, heating and service hot water provisions of the code. [Ref. R402.1]

FIGURE 12-3 Tropical climate zone lanai.

TROPICAL ZONES

A special exemption to the provisions of the IECC exists for residential buildings located in the tropical climate zone. The code for structures in this climate zone translates traditional design methods with appropriate modern technology into 11 prescriptive language requirements. Buildings in Hawaii, Puerto Rico and the U.S. Virgin Islands typically make use of provisions that recognize that the tropical climate is unusually moderate and comfortable. Buildings below 2,400 feet in elevation in this zone comply with Chapter R4 when the provisions are met. This allows for the living spaces such as the one shown in Figure 12-3.

The occupied space cannot be heated and no more than one-half can be air conditioned. In any conditioned portion of the building, glazing must have a solar heat gain coefficient (SHGC) of less than or equal to 0.40 or an overhang with a projection factor equal to or greater than 0.30. The ceiling must be insulated, or the exterior roof surface must comply with one of the solar reflectance and remittance options in Table C402.3, and roofs must have a minimum slope of ¼ inch per foot of run as in Figure 12-4.

Interior doors to bedrooms are able to be secured in the open position, and bedrooms with exterior walls facing two different directions must have operable

FIGURE 12-4 Tropical single-family home.

fenestration on exterior walls facing two directions. This provision allows tropical breezes to cross ventilate and cool the bedrooms. A ceiling fan or ceiling fan rough-in must be provided for bedrooms and the largest space that is not used as a bedroom.

Unless mechanical ventilation is present, operable fenestration needs to provide a ventilation area equal to at least 14 percent of the floor area in each room, and 80 percent of the energy for service water heating must come from a renewable energy source. Permanently installed lighting must still meet the provisions of Section R404. **[Ref. R401.2.1]**

COMPLIANCE

The compliance path for residential buildings is the same as that for commercial buildings, and will be briefly reviewed here; please see Section 3 of this text for a complete discussion. Whether dealing with new construction, alterations, repairs or an addition, the plans, specifications and details must comply with the minimum code provision. The simplest compliance path for single-family and noncomplex residential structures is to use the prescriptive worksheets offered by some jurisdictions that establish the requirements applicable to that location. Another option, discussed fully in Section 3, is the U.S. Department of Energy's (DOE's) RES*check*™ software program, which is the residential version of COM*check*™. Most, but not necessarily all, jurisdictions will accept RES*check*™ documentation. It is always best to talk with the building official before using any type of software or worksheets and determine if the product and its results are acceptable to the jurisdiction. **[Ref. R101.5.1]**

You Should Know

The ICC Evaluation Service (ICC-ES) is an independent body that evaluates and issues reports of compliance based on various codes and standards. ICC-ES Evaluation Reports (ESRs) are available to all users free of charge and provide a valuable tool for the safe and consistent use of construction materials. ICC-ES also provides evaluation of sustainability attributes, which support the building industry's research, development and manufacturing of efficient and effective energy-specific materials and components. ●

ALTERNATE MATERIALS

As noted, the intent of the code is to encourage innovative designs, materials and techniques for energy savings; thus, any material or method of construction that meets the intent of the applicable code provision may be evaluated and approved by the code official. **[Ref. R102.1]** In addition, many jurisdictions adopt, administer and enforce mandatory above code programs, which may be deemed compliant by the code official. However, as noted in Section 3, even when an alternate program is deemed to comply, the mandatory energy efficiency provisions in Chapter 4 of the IECC must be met. **[Ref. R102.1.1]**

EXISTING AND HISTORIC BUILDINGS

The code provisions for existing and historical residential buildings are the same as those for existing and historical commercial buildings and will be briefly summarized here; please see Section 3 of this text for a complete discussion of these regulations. The IECC does not require retroactive alterations to legally existing buildings or building systems to bring them into compliance with the minimum energy code provisions, but the IECC provisions do apply to buildings constructed without a permit and most types of work proposed for an existing building that require a permit. [Ref. R501.2]

Existing buildings also may be classified as currently listed, designated or certified historic, like the one shown in Figure 12-5, or in the process of certification, in which case the historic building provisions of the IECC apply. In such a case, all aspects of alteration or repair of historic buildings are exempt from the energy code. (See Section 3 for a complete discussion.) Although a historic building is not required to comply with the IECC, the owner could elect to use the guidance of the code provisions in order to make the building more energy efficient; the key distinction is that the IECC is an option rather than an enforceable requirement. [Ref. R501.6]

Code Essentials

"Historic buildings" are defined in the IECC as one or more of the following:

1. Listed, or certified as eligible for listing by the State Historic Preservation Officer or the Keeper of the National Register of Historic Places, in the National Register of Historic Places.
2. Designated as historic under an applicable state or local law.
3. Certified as a contributing resource within a National Register-listed, state-designated or locally designated historic district.

FIGURE 12-5 Designated historic single-family home.

PERMITS

The IECC does not define the specific "energy code" permits that are required. However, the IECC does define the work that must comply with the minimum code requirements and the types of work that require a permit to meet compliance. Portions of an existing building undergoing renovation or repair (but not the unaltered or unaffected portions of the building, which are exempt) must comply with energy

You Should Know

Most jurisdictions consider "legally existing" as a permitted, inspected structure that was issued a certificate of occupancy.

code provisions required for new construction. Any addition to an existing building also must comply with energy code provisions required for new construction.

Certain types of work, however, do not require compliance, such as proposed construction or alterations that will not increase building energy use. For example, storm windows installed over existing windows improve thermal performance and reduce energy use. This popular homeowner project is specifically exempted from the energy performance requirements for new windows. The code also exempts replacing broken glass in existing window frames. This exception does not, however, exempt whole window replacements. Installing a new window assembly in an existing window opening requires the window to meet the same solar heat gain coefficient (SHGC) and *U*-factor requirements as would be required for new construction. Typical energy-upgrade projects that include window replacement require permit application, plan review, permit issuance and inspection. **[Ref. R503.1]**

CHANGE IN OCCUPANCY OR USE

Changes in residential tenant occupancy seldom trigger energy code compliance. Apartment tenants and renters come and go and homes are remodeled, but the classification of "residential occupancy" is unaffected.

For residential buildings, a change in use most often happens when any unconditioned space becomes heated or cooled, and this may trigger code compliance. For example, a garage, outbuilding, storage area or basement without heating or cooling that becomes a den, bedroom, apartment or living area must comply with all the provisions of the code. Any change in use from another occupancy to residential must fully comply with the code. **[Ref. R505.2]**

All residential buildings—including multi-family residential buildings not more than three stories high—are regulated by the provisions in this part of the code. As noted in the commercial provisions discussion, in buildings not more than three stories tall that contain both commercial spaces and residential units, the residential units must comply with these code provisions. Hotels or apartment buildings four stories or more, even though the entire use is residential, must meet the commercial energy provisions. **[Ref. R101.4.1]**

FIGURE 12-6 A garage that has been remodeled for use as an apartment.

CHAPTER

13 Administration and Enforcement

PREPARING THE PLANS

As with commercial buildings, the permit applicant for residential buildings is responsible for preparing a complete set of construction documents for the building project. In many cases, a set of building plans with sufficient details and schedules will describe a simple project and demonstrate compliance with the energy code. A more complete understanding of the code provisions and requirements by the homeowner-builder, designer and contractor will help the building department plans examiner review the project for compliance and speed the process of permit issuance. It is important to determine if the jurisdiction requires the plans to be prepared by a registered design professional. The local code official will have information about the applicability of this requirement to specific projects, and is also authorized to waive this requirement if it is determined that a registered design professional's stamp and seal is not required to confirm compliance with the IECC. In most cases, approved compliance software may not be required to be submitted by a design professional. [Ref. R103.1]

Elements of the building envelope, mechanical equipment, duct sizing, and hot water pipe and duct insulation information must be included in the plans for energy code compliance review. [Ref. R103.2]

The prescriptive or performance compliance path chosen will determine the detail required to demonstrate code compliance. The prescriptive provisions provide a simple path to building document preparation.

THE CONSTRUCTION DRAWINGS AND DOCUMENTS

The thermal envelope and air sealing

As noted earlier in Section 4, the plans must include building elevations depicting window, door and skylight areas and wall sections showing the type of insulation and its thickness. A good trick to remember is to use your finger to trace the thermal envelope on the drawing sections and details (Figure 13-1). The design documents should have sufficient detail to trace all six sides of the building cube without lifting your finger off the paper. Three sides of the cube are shown in this figure. This trick also works to confirm the air-sealing continuity in the design drawings.

The wall and roof sections in the plans must list the *R*-value of the insulation (Figure 13-2), and the window and door schedules must list the *U*-factors and solar heat gain coefficient (SHGC) of the specified fenestrations. Above- and below-grade wall construction and framing must be identified separately.

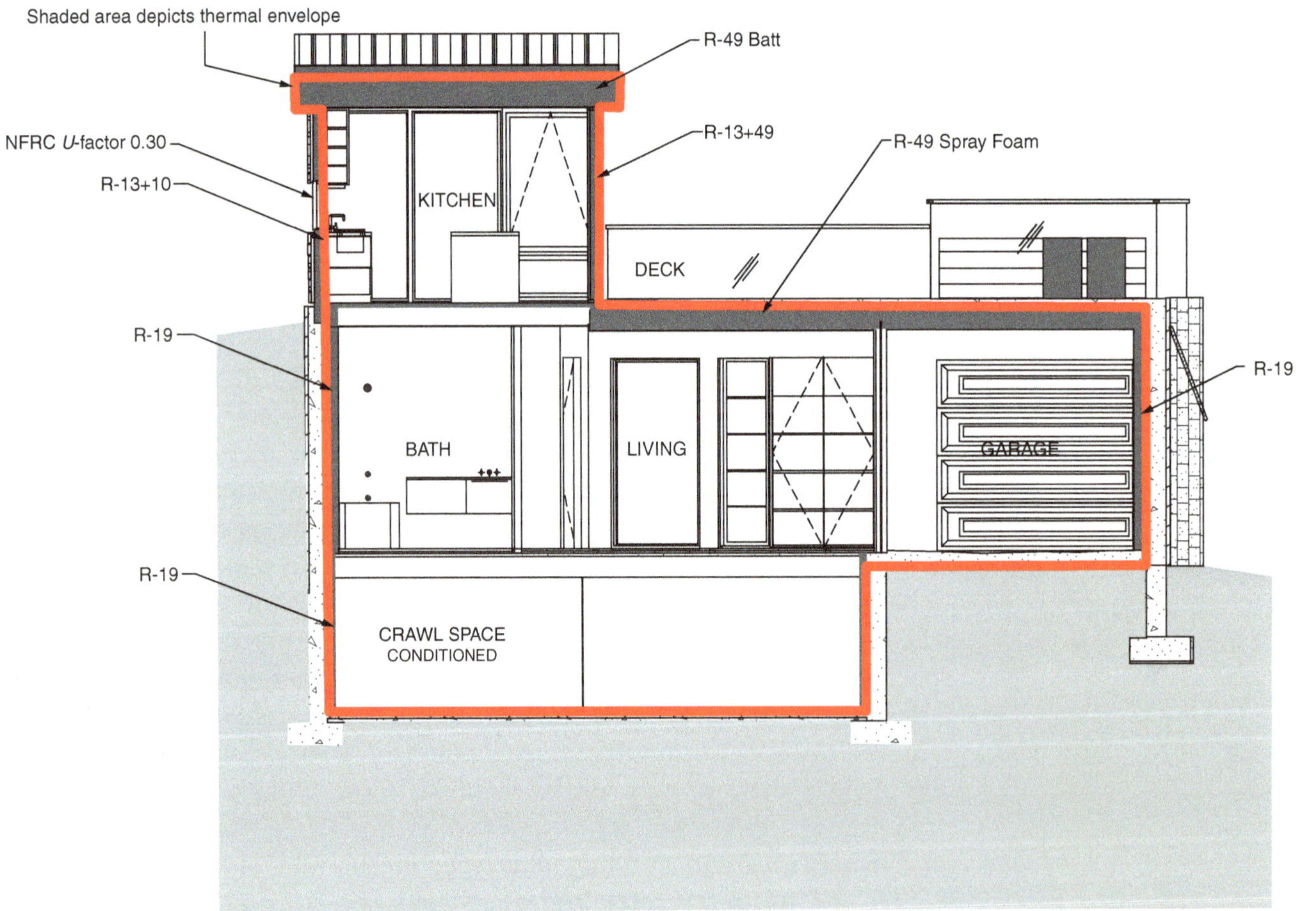

FIGURE 13-1 Thermal envelope for Climate Zone 6.

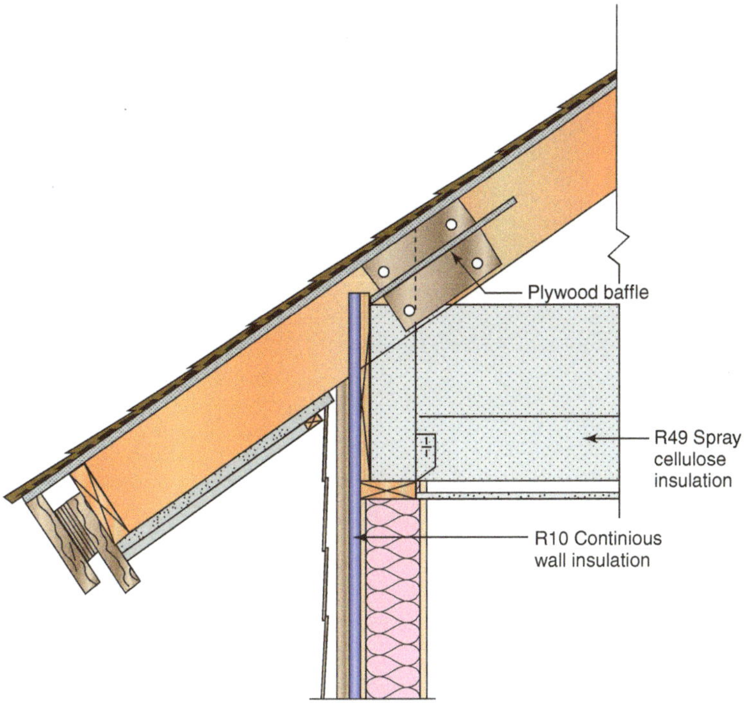

FIGURE 13-2 *R*-value of the insulation.

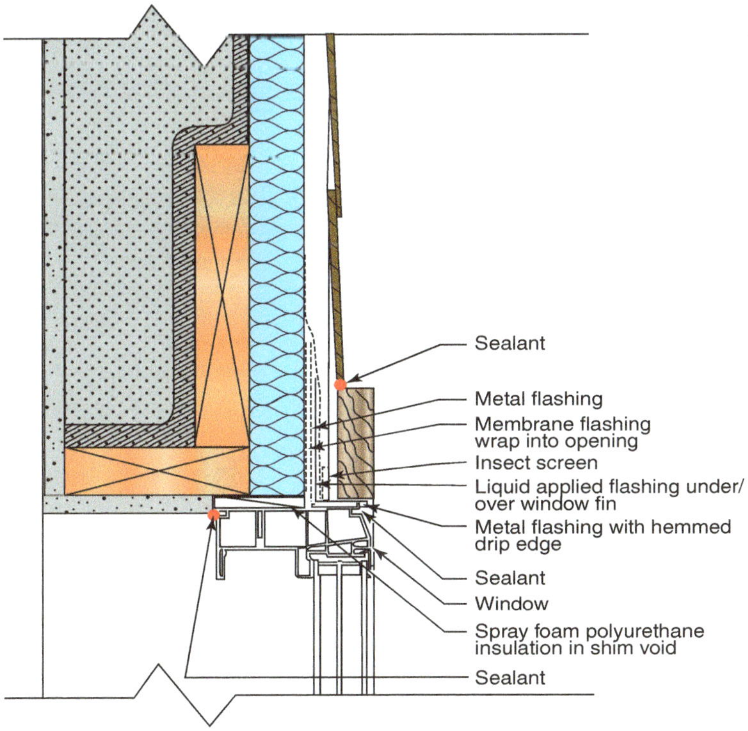

FIGURE 13-3 Details on how exterior joints should be sealed. *(Courtesy of Building Science Corporation)*

The insulation and material requirements differ for wood- versus metal-framed walls. Building plan elevations indicating window and door sizes and locations, and a roof plan showing skylights, must agree with the schedules, notes and callouts. Caulking and sealing details for window and door frame installation must be included in the construction documents submitted for plan review. Figure 13-3 shows notes and details that are typical of how all exterior joints are to be sealed.

Air infiltration and exfiltration must be addressed in the building plan documents. All exterior joints, cracks and holes are to be gasketed, caulked, weather-stripped, or sealed (Figure 13-4). The air-sealing materials must be called out on the plans and compatible with the building materials in the joints and holes where they are applied. Details, notes and sections in the building plans must address the continuity of the air sealing for the entire building envelope. **[Ref. Table R402.4.1.1]**

Building systems

Mechanical system design must be included in construction documents. Multi-family residential building; heating, ventilation and air conditioning (HVAC) and plumbing systems are sophisticated and complex, and the documents must be prepared by registered design professionals. Mechanical and plumbing systems in single-family homes may be simple compared to those for a larger residential project, and may not require preparation by a registered design professional. The documents, whether submitted by a licensed engineer or a homeowner-builder, require mechanical equipment and duct sizing calculations and HVAC schedules that include the types and efficiencies of the mechanical equipment. Duct-sealing materials must be specified to demonstrate compatibility with the duct materials. In addition, all ducts and certain hot water pipes are required to be insulated. Certain service hot water pipes are required to be insulated. High-efficacy lighting fixtures and lamps must be specified on the plans as well.

PLAN REVIEW

Just as with commercial buildings, the code official is charged with reviewing the construction documents—including schedules, calculations and specific product information—for residential buildings for code compliance. [Ref. R103.3] The final step in permit issuance is for the official to stamp the approved documents "Reviewed for Code Compliance." As with commercial projects, one set of plans is kept in the building department office and an identical set is kept at the work site. This set of construction documents cannot be altered or changed without the code official's approval. [Ref. R103.3.1]

The work site set is the reference for field inspections, and the construction must agree with the approved plans. Any changes to the approved plans must be resubmitted to the code official for further review, approved by the official and returned to the field before inspections are called for that portion of the work. [Ref. R103.4]

A moisture and vapor control strategy must be addressed in the building plan details. The requirements in Section R702.7 of the 2018 *International Residential Code*® (IRC®) (Figure 13-5) must be incorporated in the building plan's wall section details. Exterior walls must be covered with materials approved as weather coverings, and in certain climate zones they must include a vapor retarder. It is important to be familiar with and incorporate these provisions in the plans (Figure 13-6).

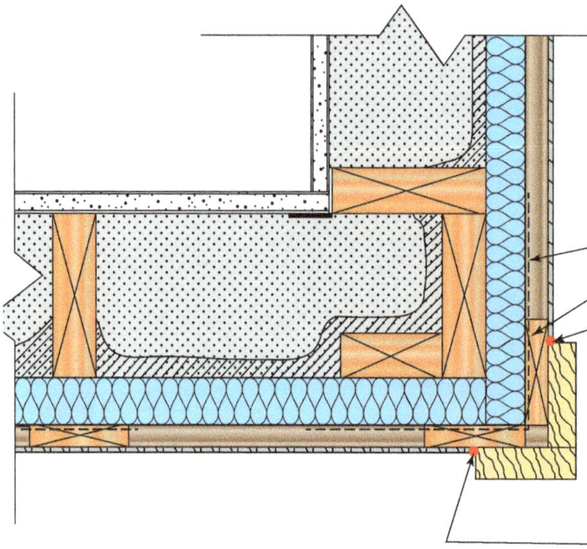

FIGURE 13-4 Exterior joint caulk and seal.

R702.7 Vapor retarders. Class I or II vapor retarders are required on the interior side of frame walls in climate zones 5, 6, 7, 8 and Marine 4.

Exceptions:

1. Basement walls.
2. Below grade portion of any wall.
3. Construction where moisture or its freezing will not damage the materials.

R702.7.1 Class III vapor retarders. Class III vapor retarders shall be permitted where any one of the conditions in Table R702.7.1 is met.

R702.7.2 Material vapor retarder class. The vapor retarder class shall be based on the manufacturer's certified testing or a tested assembly.

The following shall be deemed to meet the class specified:

Class I: Sheet polyethylene, unperforated aluminum foil.
Class II: Kraft-faced fibreglass batts.
Class III: Latex or enamel paint.

FIGURE 13-5 Moisture and vapor control must be addressed in building plan details.

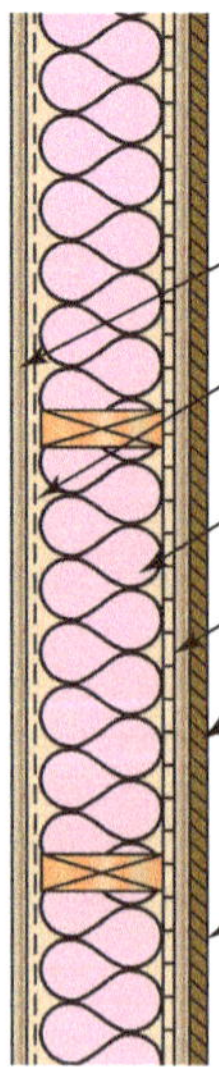

FIGURE 13-6 Vapor retarder as required.

INSPECTIONS

As with commercial projects, residential building construction that requires a permit also requires inspection. Specific energy code inspections are listed in the IECC, which complement or reiterate the inspections required in other codes. The periodic inspections required by the normal construction process will incorporate many of the energy code provisions in the inspection schedule. For example, the foundation inspection requires perimeter and below-grade wall insulation to be in place before backfilling, and fenestration compliance and air sealing are verified in the framing inspection. The insulation inspection verifies the required material *R*-values are in place and properly installed.

Mechanical and plumbing inspections are required to confirm compliance with the approved plans. The mechanical plans and schedules indicate the mechanical equipment efficiency and duct size, location and insulation. Plumbing inspections check that the equipment efficiencies and pipe insulation are installed as required and as approved on the plans. The final building inspection includes counting the number and type of high-efficacy light fixtures and lamps to verify energy code compliance. **[Ref. R105.2]** The inspections required by Section R105.2, therefore, add no additional burden onto the code official or approved inspection agency (Figure 13-7). Each installed component specific to energy compliance must perform as specified.

Code Essentials

Section 110.3.8 of the *International Building Code* (IBC), "Energy Efficiency Inspections," applies to residential occupancies other than IRC buildings, and requires compliance for those constructed under the provisions of the IBC. The inspections include, but are not limited to, envelope insulation *R*-value, fenestration *U*-value, duct system *R*-value, and HVAC and water-heating-equipment efficiency. It is the "shall include, but not be limited to" language in this inspection requirement that requires special attention.

Most state and local jurisdictions amend the codes during the adoption process; thus, it is necessary to consult the authority having jurisdiction to find out exactly what the inspection requirements are to confirm compliance with the energy code. ●

Ø **R104.2 Required inspections.** The code official or his designated agent, upon notification, shall make the inspections set forth in Sections R104.2.1 through R104.2.5.

Ø **R104.2.1 Footing and foundation inspection.** Inspections associated with footings and foundations shall verify compliance with the code as to R-value, location, thickness, depth of burial and protection of insulation as required by the code and approved plans and specifications.

Ø **R104.2.2 Framing and rough-in inspection.** Inspections at framing and rough-in shall be made before application of interior finish and shall verify compliance with the code as to types of insulation and corresponding R-values and their correct location and proper installation; fenestration properties (U-factor and SHGC) and proper installation; and air leakage controls as required by the code and approved plans and specifications.

Ø **R104.2.3 Plumbing rough-in inspection.** Inspections at plumbing rough-in shall verify compliance as required by the code and approved plans and specifications as to types of insulation and corresponding R-values and protection, and required control.

Ø **R104.2.4 Mechanical rough-in inspection.** Inspections at mechanical rough-in shall verify compliance as required by the code and approved plans and specifications as to installed HVAC equipment type and size, required controls, system insulation and corresponding R-value, system air leakage control, programmable thermostats, dampers, whole-house ventilation, and minimum fan efficiency.

Exception: Systems serving multiple dwelling units shall be inspected in accordance with Section C104.3.4.

FIGURE 13-7 Required residential energy code inspections.

FEES

As covered in Chapter 4, jurisdictions establish a fee schedule when they adopt code provisions. Fee amounts vary, so specific information must be gathered from the jurisdiction responsible for the proposed project. Any and all applicable fees must be paid before the permit is issued. Energy code compliance may require building, mechanical, plumbing and electrical permits. Any work that needs a permit should not begin before the applicable permit is issued. **[Ref. R104.1]** Starting work without a permit may trigger additional fees to be added to the permit fee. **[Ref. R104.3]** It is not a good idea to start work before a permit is issued because it may not comply with the applicable code provisions—resulting in more time, money and materials than necessary.

ENFORCEMENT

A stop work order is the ultimate enforcement tool. The code official has the authority to stop work on any part of a project if the work regulated by the code is dangerous, unsafe or not code compliant. **[Ref. R108.1]** The stop work order is issued in writing to the owner or person doing the work and is specific to the noncompliant work that must be stopped. **[Ref. R108.2]** If the stop work order is ignored, the code official may levy a fine, an action that generally involves a citation in municipal court. The amount of the fine is set by each jurisdiction as determined at the time of code adoption. **[Ref. R108.4]**

BOARD OF APPEALS

As with commercial construction, the board of appeals exists to allow a means to appeal decisions made by the building official concerning residential construction. Members of appeal boards are usually civic-minded architects, engineers, insulators, lighting designers or mechanical, plumbing or electrical contractors. **[Ref. R109.3]** The board of appeals decides whether the appellant or the code official's action is consistent with the intent of the code. In some cases, the energy code may not apply to a specific project condition; in others, an alternative approach to a proposed construction method may be rejected by the code official because it does not appear in the code, but the board may deem that the method or material is equal to that described in the code.

Note that it is the *intent* of a certain code provision that generates the basis for an appeal. An appeal can be heard when an applicant, designer or tradesperson does not agree with the code official's decision or order. The appeal may concern a design or field inspection decision and be based on an opinion that the true intent of the code provision or rule has been incorrectly applied by the code official. **[Ref. R109.1]** The board hears the appeal, makes a decision and records the findings in writing. In

Code Essentials

"Zone": A space or group of spaces within a *building* with heating or cooling requirements that are sufficiently similar so that desired conditions can be maintained throughout using a single controlling device. ●

no case may the board waive any requirements of the code. [Ref. R109.2]

DEFINED TERMS

The 70 specific words and terms defined in Chapter 2 [RE] of the IECC have a common and agreed-on meaning when written and appearing in the code provisions. [Ref. R201] For example, the words *accessible*, *readily accessible*, *building*, *listed*, *repair*, *basement wall* and *zone* all are defined. Any defined word or term appears in *italics* in the code text. In reading, interpreting and applying code provisions, it is essential to use the common code vocabulary. [Ref. R202] This is the basis for civil discussion and deeper understanding of the energy code provisions.

PART

V

Specific Requirements for Residential Buildings

CHAPTER

14 Specific Requirements for Residential Buildings

GENERAL

An Energy Efficiency Certificate listing installed insulation, glass and door ratings, and heating and cooling equipment efficiency and test results must be posted on a wall in the space where the furnace is located, in a utility room or in an approved location inside the building. The certificate is useful consumer information comparable to the option package and estimated miles-per-gallon sticker on new car windows. It creates a permanent record of building elements that are difficult to determine after the construction is finished.

The certificate must include building envelope, duct insulation, and window and door *U*-factor and solar heat gain coefficient (SHGC) values. The window and door information is valuable when replacement projects are contemplated. The Heating, ventilation, and air conditioning (HVAC) system and water heater efficiencies are required to be listed. This information is a guide to maintaining or increasing energy savings when any of the equipment is replaced. The blower door and duct leakage test results are also recorded. The whole house number establishes the baseline with which to compare future test results to evaluate air barrier performance over time.

One example of residential energy compliance software that has the option to generate a certificate (see Figure 14-1) in addition to a compliance report is Res*check*™. The program cannot list out every assembly on the certificate. The information for insulation rating and glass and door rating is automatically populated with the area-weighted averages. For example, if there were windows that had different *U*-factors, the area-weighted average would be displayed. Furthermore, additional information will need to be filled out manually (water heater efficiency, duct insulation, air and duct leakage test results). **[Ref. R401.3]**

The certificate must include building envelope, duct insulation, and window and door *U*-factor and SHGC values. The window and door information is valuable when replacement projects are contemplated. HVAC system and water heater efficiencies are required to be listed. This information is a guide to maintaining or increasing energy savings when any of the equipment is replaced. The blower door and duct leakage test results are also recorded. The whole house number establishes the baseline with which to compare future test results to evaluate air barrier performance over time.

2015 IECC Energy Efficiency Certificate

Insulation rating	R-Value	
Ceiling/Roof	59.00	
Wall	30.00	
Floor/Foundation	15.00	
Ductwork (unconditioned spaces):	______	
Glass & door rating	**U-Factor**	**SHGC**
Window	0.28	0.40
Door	0.35	NA
Heating & cooling equipment	**Efficiency**	
Heating system: ______________	______	
Cooling system: ______________	______	
Water heater: ______________	______	
Building air leakage and duct test results		
Building air leakage test results	______________	
Name of air leakage tester	______________	
Duct tightness test results	______________	
Name of duct tester	______________	

Name: ______________________ Date: __________

Comments:

FIGURE 14-1 RES*check*™ Energy Efficiency Certificate Example.

CHAPTER 15

Building Thermal Envelope Requirements

ABOUT THE THERMAL ENVELOPE

The building thermal envelope separates the exterior from the interior conditioned space. Building elements of the thermal envelope include basement and exterior walls, roofs, windows, doors, skylights and floors. The key to establishing the boundaries of the thermal envelope is found in the definition of *"Conditioned space."* As defined in the *International Energy Conservation Code*® (IECC®), *"Conditioned space"* is "an area, room or space that is enclosed within the *building thermal envelope* and that is directly or indirectly heated or cooled. Spaces are indirectly heated or cooled where they communicate through openings with conditioned spaces, where they are separated from conditioned spaces by uninsulated walls, floors or ceilings, or where they contain uninsulated ducts, piping or other sources of heating or cooling." It is important that the project designer defines the limits of the building thermal envelope in the plans and demonstrates compliance with the mandatory and applicable requirements of Chapter 4 of the IECC. In some designs, crawl spaces, attics and areas behind knee walls are unconditioned and exposed directly to the outdoors, separated only by uninsulated exterior walls or roofs. In such cases, the building thermal envelope separates these unconditioned spaces from the interior of the building.

CODE COMPLIANCE PATHS

The code offers four choices for compliance with the building thermal envelope requirements: Prescriptive, Total UA Alternative, Performance and Energy Rating Index (ERI). Table R402.1.2 offers a very simple and straightforward compliance method. Individual components of a typical building envelope are listed and the performance level of each is established by climate zone. Not every building design will use every component, and it is important to be familiar with the definitions and descriptions of these components. Only then can the required minimum

You Should Know

The U.S. Census Bureau reports that in 2016, there were 135,697,926 existing housing units in the United States.[1] Also in 2016, 1,206,643 residential building permits were issued. The census data summarized in Figure 15-1 show that about 60 percent of these homes were built before 1980. The wall insulation values for these existing homes are not separated by climate zones but are well below values in the current energy code provisions[2] (Figure 15-1). The energy code provisions require areas in a building undergoing an alteration, renovation or repair to conform to the provisions for new construction. The ultimate intent of this code requirement is that over time, with few exceptions, the wall insulation in existing homes and buildings will conform to the current energy code provisions. **[Ref. R501]** •

[1] U.S. Bureau of the Census. 2017. Census of Population and Housing, Profiles of General Demographic Characteristics. http://www.census.gov/quickfacts.
[2] "Exterior Wall Insulation." Madison Gas & Electric. http://tinyurl.com/MGE-Exterior.

Year of construction	Percent of total housing units*	Wall insulative R-value
Before 1950	20.40%	4
1950–1960	11.50%	3 or 6.2
1960–1970	11.60%	6.2 or 11
1970–1980	16.70%	11 and 5†
1980 to present	39.90%	11 and 5;† 19 and 5†,‡

* There are roughly 127.7 million homes in the U.S., according to data taken from 2005–2009.
† "and" indicates use of multiple types of insulation.
‡ Each set of numbers resents the R-value for a common insulation combination used.

Table 1: U.S. housing stock and typical insulation R-values per construction year.

Window type	Annual cost
Single-glazed with storm window	$1,310
Double glazed	$1,228
Double glazed with low-e	$1,120
No window (Insulated walls only)	$1,000
Super window	$960

Table 2: Heating costs (average house in a heating climate) with different window types.

FIGURE 15-1 Fenestration and wall insulation.

R-values for the ceiling and insulation components for the wood-frame, basement or crawl space walls and *U*-factor and solar heat gain coefficient (SHGC) values for windows and skylights be determined. Lower-level house designs may include combinations of all of these elements. When building components are clearly identified in the drawings, the minimum *R*-values can be specified for the required insulation and the *U*-factor and SHGC values for the windows and skylights.

Only the *R*-value of the insulation products (Table 15-1) is used to confirm compliance with the requirements of this table; no consideration is given for any component other than the insulation products to determine the *R*-value of the opaque assemblies in the thermal envelope component. This illustrates the importance of the insulation certificate.

TABLE 15-1 Insulation *R*-value primer table.

Fiberglass batt insulation characteristics	
Thickness (inches)	Typical *R*-value
3 ½	11
3 ⅝	13
3 ½ (high density)	15
6 to 6 ¼	19
5 ¼ (high density)	21
8 to 8 ½	25
8 (high density)	30
9 ½ (standard)	30
12	38
Foam plastic insulation characteristics	
Polystyrene	R-3.8 to R-5 per inch
Polyisocyanurate	R-5.6 to R-8 per inch
Polyurethane	R-7 to R-8 per inch
Cellulose is a loose-fill insulation material that provides a thermal resistance of R-3.6 to R-3.8 per inch.	

The footnotes to this table offer design choices and specific direction regarding insulation placement. Footnote h allows wood-frame walls in Climate Zones 3 through 5 to be constructed with R-20 cavity insulation or R-13 cavity insulation plus R-5 continuous insulation. This allows the designer or builder to choose 2 × 6 or 2 × 4 stud wall construction. **[Ref. Table R402.1.2]**

Table R402.1.4 is the *U*-factor version of Table R402.1.2 and offers the *U*-factor computation method in place of the *R*-values. All parts of the construction assembly (Figure 15-2) that contribute to lessening thermal transfer are given a value and considered in the *U*-factor computation. This method is advantageous when using construction methods that limit framing (24" OC versus 16" OC) or include thermal breaks. Resource material outside of the IECC must be consulted to determine compliance with this method. The ASHRAE *Handbook of Fundamentals* is often used to calculate the *U*-factor.

Section R402.1.5 offers a tradeoff path to compliance. This UA alternative allows a design for building components that do not comply with the requirements of Table R402.1.4, as long as the area-weighted average thermal resistance is at least equivalent. Calculations must demonstrate that the total UA is the same as or less than that of a building of the same design that complies with the equivalent *U*-factor method in Table R402.1.4 (Figure 15-3). Supporting calculations simply multiply the area of the component (walls, roof, windows and doors) by the proposed design *U*-factor. The same component area is multiplied by the *U*-factor required in Table R402.1.4. The sum of the proposed design must be less than or equal to the standard code design for compliance.

The simulated performance alternative is the fourth path to energy code compliance and offers flexibility to determine the most cost-effective methods to design a code-compliant structure. This method uses software to predict the annual energy use of the proposed design compared to that of the standard design of building components listed in Table R405.5.2(1). If the proposed building's annual energy use is less than or equal to that of the standard design, the building complies with the code's goal of effective energy use. This method is useful to consider for any building designs that incorporate shading strategies to minimize

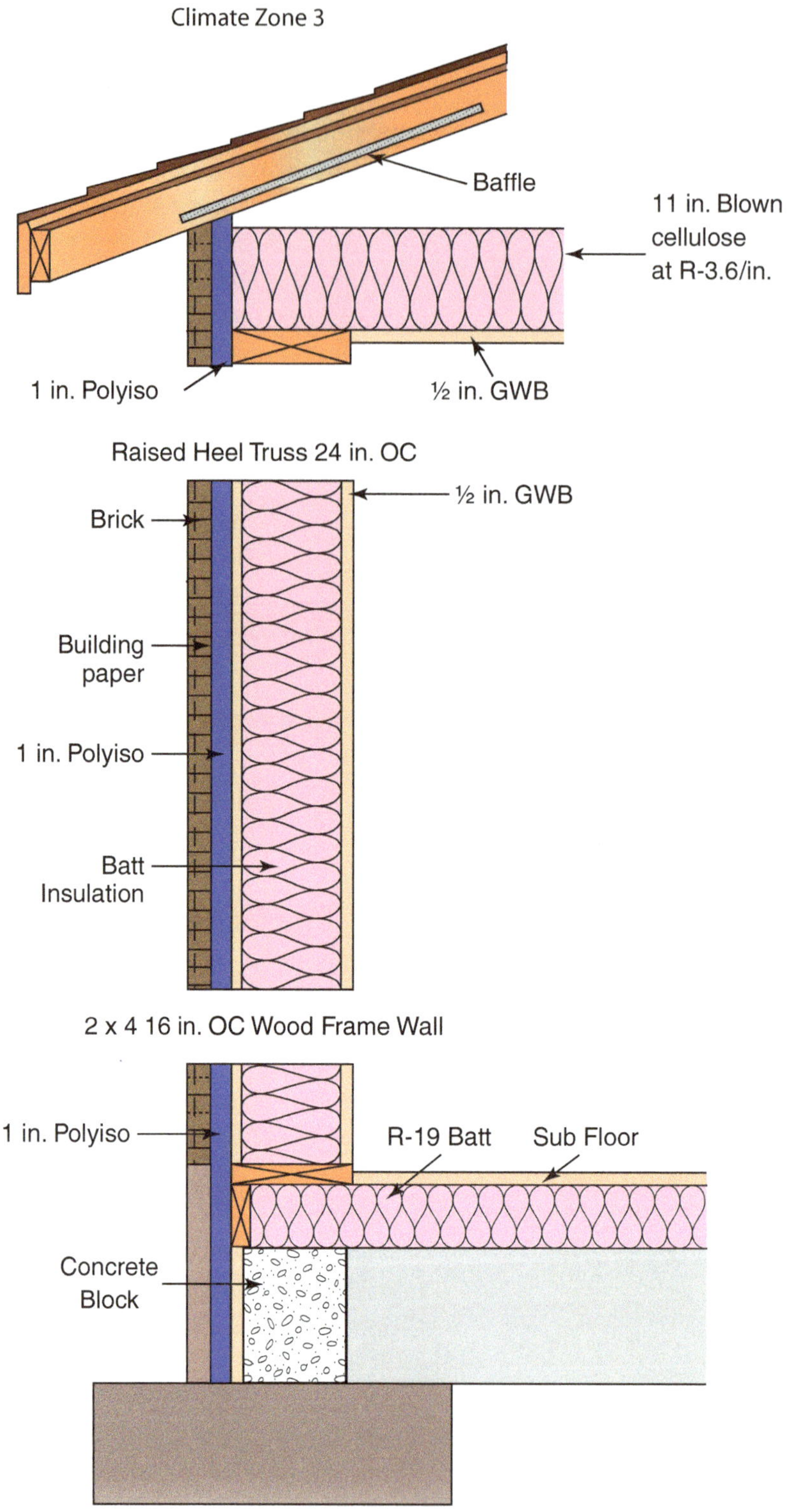

FIGURE 15-2 Building thermal envelope assemblies.

Climate zone (301.1): 7

Climate zone (301.1): 7

Ceilings and floors over outside air

Description	Insulation R-value	U-factor	x Area	= UA
Ceilings	60	0.017	800 ft²	13.3
			ft²	
Floors over outside air			ft²	
		Ceilings: Total area	800 ft²	

Required U-factor	x Area	= UA
0.026	800 ft²	20.8

Walls

Description	Insulation R-value	U-factor	x Area	= UA
Walls	19	0.0526	1080 ft²	56.84
			ft²	
		Walls: Total area	1080 ft²	

Required U-factor	x Area	= UA
0.057	1080 ft²	61.56

Vertical fenestration (Windows and doors)

Description	Insulation R-value	U-factor	x Area	= UA
Windows	—	0.33	78 ft²	25.74
	—		ft²	
Doors	—	0.40	42 ft²	16.8
	—		ft²	
Sliding glass doors	—		ft²	
	—		ft²	
		Vertical fenestration: Total area	120 ft²	

Required U-factor	x Area	= UA
0.32	120 ft²	38.4

Floors and foundations

Description	Insulation Depth	Insulation R-value	U-factor	Area or x Perimeter	= UA
Floors over Uncond. space	NA			ft²	
Basement walls	NA			ft²	
Unheated slabs	ft.			ft	
Heated slabs	4 ft.	10	0.1	120 ft	12
Crawl space walls	ft.			ft²	
				ft²	
				Total proposed UA	124.68

Required U-factor	Area or x Perimeter	= UA
0.67	120 ft²	8
	ft²	
	ft	
	ft	
	ft²	
	ft²	
	Total required UA	128.76

FIGURE 15-3 UA alternative computation.

You Should Know

The 2013 American Society of Heating, Refrigerating and Air-Conditioning Engineers, Inc. (ASHRAE) *Handbook of Fundamentals* is a standard referenced in Chapter 6 of the IECC. **[Ref. R402.1.5, Table 405.5.2(1)]** The handbook covers basic principles and includes data to support complete simple and complex calculations to demonstrate energy code compliance. Chapter 26, "Heat, Air, and Moisture Control in Building Assemblies—Material Properties," includes *R*-values for common building and insulating materials in Table 4. Values referenced in this table are used to determine equivalent *U*-factor or total UA (thermal transmittance) alternative compliance. ●

air-conditioning loads in the summer and maximize heat gain in the winter. Window orientation, efficient mechanical systems and on-site renewable energy sources are allowed input values in this compliance method. **[Ref. R405]**

RES*check*™ is a free software program available from the U.S. Department of Energy (DOE). It can be used to evaluate performance design to show compliance with the provisions of Section R405. RES*check* can be used to demonstrate that the proposed building meets the efficiency requirements of the code by choosing "Performance Alternative" in the Compliance Method menu. Compliance with Section R405 does not eliminate the mandatory requirements that are referenced in Section R401.2. As an example, the air leakage provisions of Section R402.4 are still applicable even though the software calculations or performance alternative may indicate that the building is energy efficient. This is important information and is easy to find in the code book because the mandatory requirements are marked as a reminder.

You Should Know

The U.S. Department of Energy offers free software to determine code compliance. The latest software version of RES*check* is updated to the 2018 IECC. ●

The Energy Rating Index (ERI) is another path that can be used to demonstrate compliance with the energy code. The ERI is an Index scoring system with a number based on the ERI reference design value of 100. A residential building that uses no net purchased energy has an Index value of 0. The ERI reference design is configured such that it meets the minimum requirements of the 2006 IECC. Each number on the scale represents a 1 percent change in the total energy use of the rated design relative to the total energy use of the ERI reference design benchmarked at 100 on the scale. The ERI score considers all energy used in the residential building. Calculate the ERI Reference Design Ventilation rate using Equation 4-1. **[Ref. R406.3]**

TABLE 15-2 Table R406.4: Maximum energy rating index.

Climate zone	ERI
1	57
2	57
3	57
4	62
5	61
6	61
7	58
8	58

Example:

What is the ERI Reference Design Ventilation rate for an 1,850 square-foot three-bedroom home?

$(0.01 \times 1{,}850) + [7.5 \times (3 + 1)]$
$(18.5) + (7.5 \times 4)$
$18.5 + 30 = 48.5$ CFM

CFM = (0.01 × "total square foot area of house") + [7.5 × ("number of bedrooms" + 1)]

Compliance based on an ERI analysis requires that the rated design be shown to have an ERI less than or equal to the appropriate value listed in Table R406.4, based on the climate zone in which the building is located (see Table 15-2 above).

Like all pathways, the ERI path must demonstrate compliance with all mandatory requirements of the code. In addition, the ERI path must demonstrate compliance with hot water pipe insulation requirements of the 2006 IECC, and building thermal envelope and glazing requirements of the 2009 IECC if no solar has been installed on the building. If solar has been installed on the building, then the minimum thermal envelope and glazing requirements of the 2015 IECC must be installed. This thermal envelope "backstop" was designed in the code to ensure that components such as solar that can greatly lower the ERI score do not override mandatory envelope efficiency measures that are required in other pathways.

Verification of compliance with this method must be completed by an approved third party, and documentation including compliance reports must be provided to the code official. **[Ref. R406.5]** The ERI index score and the software tools used for determining the ERI shall be approved in accordance with the ANSI/RESNET/ICC 301 standard. ERI reports must verify compliance, identify the building and include an inspection list specific to the project. It will also include the name and version of the compliance software tool and identify the person who prepared the reports. **[Ref. R406.6]**

CEILINGS AND ATTICS

Table R402.1.2 provides the insulation and fenestration requirements; however, Section R402.2 allows two specific insulation reductions. When framing or truss design allows the ceiling insulation to extend over the top of the exterior wall plate, the insulation *R*-value can be decreased from R-38 to R-30 or from R-49 to R-38. The truss design is specified and commonly called out as a "raised-heel" or "energy truss." This framing and insulation technique adds additional insulation above the exterior wall reducing heat loss adjacent to the roof eves and is often used in cold climates to reduce the occurrence of ice damming and icicles. The proper air barrier at the junction of the top plate and the top of the exterior walls is required and important in areas subject to freeze/thaw cycles. **[Ref. Table R402.4.1.1(Walls)]** Another reduction is offered for cathedral ceilings or ceilings without attic spaces. This construction technique creates bigger-feeling interior spaces by applying a gypsum wallboard (GWB) ceiling finish directly to the underside of the sloped roof rafter. This limits the available insulation to the depth of the roof or ceiling framing member. In climate zones where ceiling insulation is required to be R-38 or R-49, not more than 500 square feet of the ceiling area can be insulated to R-30 when the framing member depth will not accommodate the insulation required by Table R402.1.2. **[Ref. R402.2.2]**

These exceptions are examples of apparent code requirement conflicts. The ceiling insulation requirements are reduced by specific conditions. Although Table R402.1.2 specifies an insulation level, the ceiling insulation requirements complying with these specific conditions are reduced. Figure 15-4 illustrates a method of combining insulation materials to meet requirements.

FIGURE 15-4 Spray foam and batt insulated ceiling.

Attic ventilation

Attic ventilation is required by Section 1202.2 of the *International Building Code®* (IBC®) and Section R806 of the *International Residential Code®* (IRC®). The insulation must be held back and not block the vent area. A baffle needs to be installed next to eave or soffit vents and maintain an open space the same size as, or larger than, the size of the vent. **[Ref. R402.2.3]** This detail should be included in the plans at permit submittal. IRC 806.5 includes provisions for unvented attic or rafter spaces.

Attic access is required by IBC 1208.2 and IRC R807. The hatches or doors that connect conditioned and unconditioned spaces must be insulated to the same or greater *R*-value as the area around them. **[Ref. 402.2.4]** This maintains the integrity of the thermal envelope. These are often site built and do not require special methods or materials (Figure 15-5).

The IRC and IBC do not reference the energy code in these respective sections; however, the provisions for eave baffles and attics and crawl spaces must be included in the building plan documents and details.

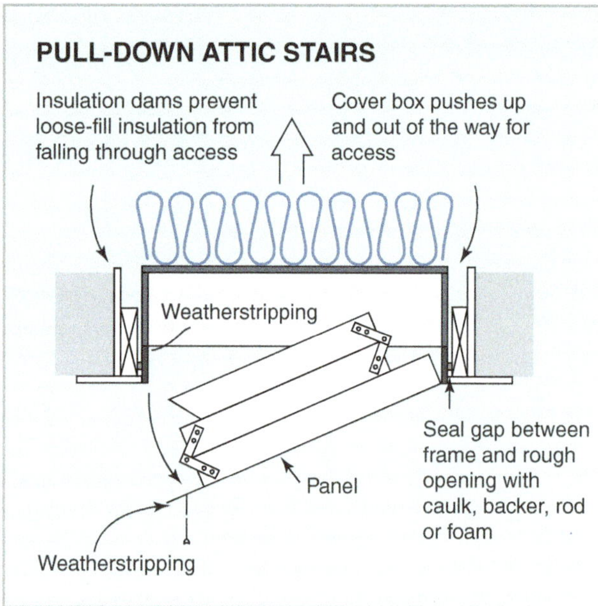

FIGURE 15-5 Attic pull-down stairs. *(Courtesy of EnergySavers)*

WALLS

The energy code recognizes a number of different types of walls in Section R402.2. This is a necessary distinction because each material in the assembly behaves differently in the thermal envelope and because some walls are above grade, below grade (Figure 15-6) or both. A typical walkout basement or garden-level apartment has wood or steel-frame walls, basement walls and perhaps a crawl space wall. The insulation values and often the insulation type and location are different for each. This makes proper identification of each wall type extremely important, and the code relies on designers, builders, plan examiners and building inspectors to understand the definitions and distinctions.

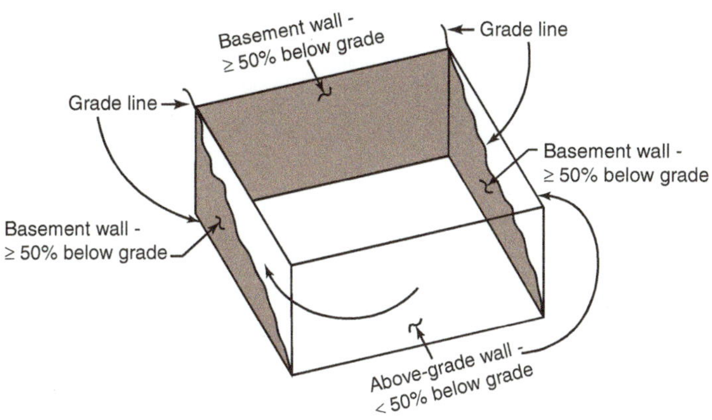

FIGURE 15-6 Typical above-/below-grade wall construction.

Code Essentials

Wall construction on a sloping grade typically requires the below-grade portions to use materials designed for basement wall construction and the above-grade portions to use materials for exterior walls. The "stepped wall" must be evaluated as either a basement or exterior wall by measuring the percentage of the wall below grade; 50 percent or more wall surface below grade defines a basement wall. Each part of the wall must meet the insulation requirements for the defined wall type. The same basement wall provisions apply to the above-grade wall portion of construction, even if it is built with materials typical of exterior wall construction.

The definitions for the various wall types are scattered around Chapter 2 of the IECC and in no particular order. Remember that when a word or term in the code is defined in Chapter 2, it is *italicized*. In alphabetical order, *above-grade wall*, *basement wall*, *crawl space wall* (do not get confused with *curtain wall*) and *exterior wall* are specifically defined. This leaves a few types of walls undefined, but careful reading of the definitions can establish the proper application of the appropriate energy code requirement. **[Ref. R202]**

Mass walls

Mass walls are built of dense materials such as concrete block, reinforced concrete (Figure 15-7), brick and insulated concrete forms (ICFs). Traditional building materials and methods such as adobe (Figure 15-14), solid logs and rammed earth are also mass wall materials. Frame walls and mass walls are very different in both performance and the amount of insulation required. Footnotes to both prescriptive tables require insulation values that are dependent on the location of the insulation (Table 15-3, Table 15-4). It makes a difference as to whether more than half of the insulation is on the interior of the mass wall. In all climate zones, more insulation is required when more than half of the insulation is on the interior of the mass wall. **[Ref. R402.2.5]**

FIGURE 15-7 Concrete mass wall.

FIGURE 15-8 t-0 Metals have a much higher thermal conductivity than wood. (Photos courtesy of Joseph Kistner).

FIGURE 15-9 t-60 Metals have a much higher thermal conductivity than wood. (Photos courtesy of Joseph Kistner).

FIGURE 15-10 t-120 Metals have a much higher thermal conductivity than wood. (Photos courtesy of Joseph Kistner).

FIGURE 15-11 t-240 Metals have a much higher thermal conductivity than wood. (Photos courtesy of Joseph Kistner).

FIGURE 15-12 t-300 Metals have a much higher thermal conductivity than wood. (Photos courtesy of Joseph Kistner).

FIGURE 15-13 t-310 Metals have a much higher thermal conductivity than wood. (Photos courtesy of Joseph Kistner).

FIGURE 15-14 Adobe wall.

TABLE 15-3 Table R402.1.2: Insulation requirements: Mass walls.

Climate zone	Mass wall *R*-value[i]
1	3/4
2	4/6
3	8/13
4 except Marine	8/13
5 and Marine 4	13/17
6	15/20
7 and 8	19/21

[i] The second *R*-value applies when more than half the insulation is on the interior of the mass wall

TABLE 15-4 Table R402.1.4: Equivalent *U*-factors.

Climate zone	Mass wall *U*-factor[b]
1	0.197
2	0.165
3	0.098
4 except Marine	0.098
5 and Marine 4	0.082
6	0.060
7 and 8	0.057

b. Where more than half the insulation is on the interior, the mass wall *U*-factors shall not exceed 0.17 in Climate Zone 1, 0.14 in Climate Zone 2, 0.12 in Climate Zone 3, 0.087 in Climate Zone 4 except Marine, 0.065 in Climate Zone 5 and Marine 4, and 0.057 in Climate Zones 6 through 8.

FIGURE 15-15 Steel framing.

TABLE 15-5 Table R402.2.6: Steel-frame ceiling, wall and floor insulation (*R*-value).

Wood frame *R*-value requirement	Cold-formed steel-frame equivalent *R*-value[a]
	Steel truss ceilings[b]
R-30	R-38 or R-30 + 3 or R-26 + 5
R-38	R-49 or R-38 + 3
R-49	R-38 + 5
	Steel joist ceilings[b]
R-30	R-38 in 2 × 4 or 2 × 6 or 2 × 8 R-49 in any framing
R-38	R-49 in 2 × 4 or 2 × 6 or 2 × 8 or 2 × 10
	Steel-framed wall 16" o.c.
R-13	R-13 + 4.2 or R-21 + 2.8 or R-0 + 9.3 or R-15 + 3.8 or R-21 + 3.1
R-13 + 3	R-0 + 11.2 or R-13 + 6.1 or R-15 + 5.7 or R-19 + 5.0 or R-21 + 4.7
R-20	R-0 + 14.0 or R-13 + 8.9 or R-15 + 8.5 or R-19 + 7.8 or R-19 + 6.2 or R-21 + 7.5
R-20 + 5	R-13 + 12.7 or R-15 + 12.3 or R-19 + 11.6 or R-21 + 11.3 or R-25 + 10.9
R-21	R-0 + 14.6 or R-13 + 9.5 or R-15 + 9.1 or R-19 + 8.4 or R-21 + 8.1 or R-25 + 7.7
	Steel-framed wall 24" O.C.
R-13	R-0 + 9.3 or R-13 + 3.0 or R-15 + 2.4
R-13 + 3	R-0 + 11.2 or R-13 + 4.9 or R-15 + 4.3 or R-19 + 3.5 or R-21 + 3.1
R-20	R-0 + 14.0 or R-13 + 7.7 or R-15 + 7.1 or R-19 + 6.3 or R-21 + 5.9
R-20 + 5	R-13 + 11.5 or R-15 + 10.9 or R-19 + 10.1 or R-21 + 9.7 or R-25 + 9.1
R-21	R-0 + 14.6 or R-13 + 8.3 or R-15 + 7.7 or R-19 + 6.9 or R-21 + 6.5 or R-25 + 5.9
	Steel joist floor
R-13	R-19 in 2 × 6, or R-19 + 6 in 2 × 8 or 2 × 10
R-19	R-19 + 6 in 2 × 6, or R-19 + 12 in 2 × 8 or 2 × 10

[a] Cavity insulation *R*-value is listed first, followed by continuous insulation *R*-value.

[b] Insulation exceeding the height of the framing shall cover the framing.

Steel-frame walls

Steel framing is sometimes specified and is common in multi-family buildings (Figure 15-15); it must comply with the residential energy code provisions. **[Ref. R402.2.6]** Steel has a much higher thermal conductivity than wood. Figures 15-8 through Figure 15-13 show how ice melts much faster in contact with steel than it does with wood of the same temperature. To adjust for the thermal transfer properties of steel, Table R402.2.6 (Table 15-5) offers steel-frame insulation combinations equivalent to the wood-frame *R*-value requirements. Choices for the combination of cavity and continuous insulation offer cost and constructability options for the designer and builder choosing steel framing in residential buildings.

Basement walls

In conditioned basements, walls (see definitions of *"Basement wall"* and *"Above-grade wall"*) **[Ref. R202]** must be insulated from the top of the basement wall to at least 10 feet below grade or to the basement floor, whichever is less (Figure 15-16). It is very important for the designer and builder to clearly indicate the final grade on the plans. This will speed up building department review for compliance and permit issuance. If the basement is designed as unconditioned space, the thermal envelope boundary must be shifted to the underside of the floor above (Figure 15-17). In this condition, the rim joist is considered to be part of the thermal envelope, and care must be taken to insulate it to the same requirements as for the floors. **[Ref. R402.2.9]**

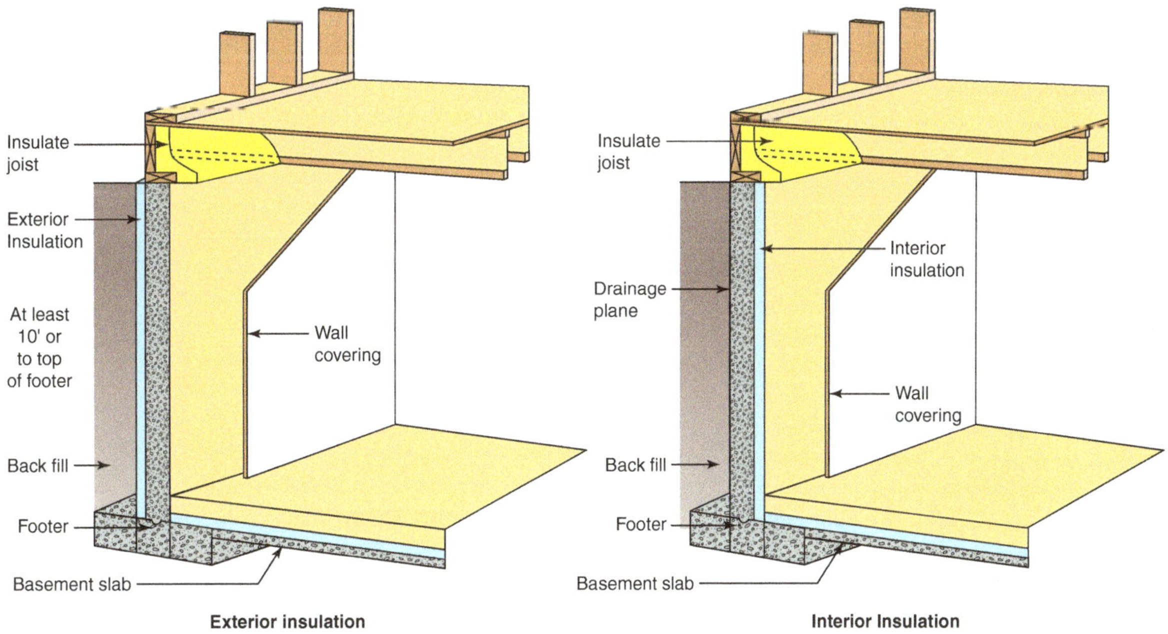

FIGURE 15-16 Basement wall insulation.

Crawl space walls

Conditioned space must be surrounded by a contiguous building thermal envelope. The code allows for the thermal envelope with respect to the crawl space to be horizontal in the floor above or vertical down the walls to create a conditioned crawl space. Unvented or vertically insulated crawl spaces are allowed in IRC Section R408.3 and IBC Section 1202.4.2. Unvented crawl spaces are popular in colder climates to prevent the freezing of pipes. In all climates, a Class I vapor retarder, typically sheet polyethylene, must cover the exposed dirt in the crawl space. The horizontal overlap joints in the vapor retarder membrane must be taped or sealed and the retarder must extend vertically 6 inches up and be sealed and attached to the stem wall (Figure 15-18).

The crawl space insulation must cover the entire wall from the floor above to the grade below and then extend horizontally or vertically at least 24 inches down the stem wall. The insulation below the crawl space floor next to the stem wall partially eliminates the thermal bridging effect created at the intersection of the foundation wall and crawl space floor (Figure 15-19). **[Ref. R402.2.11]**

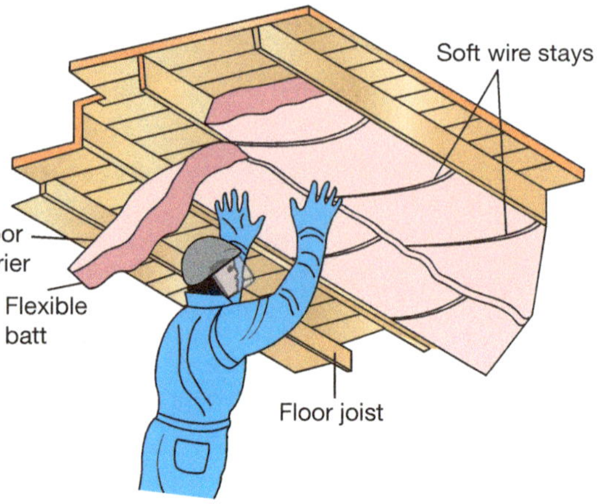

FIGURE 15-17 Basement ceiling insulation.

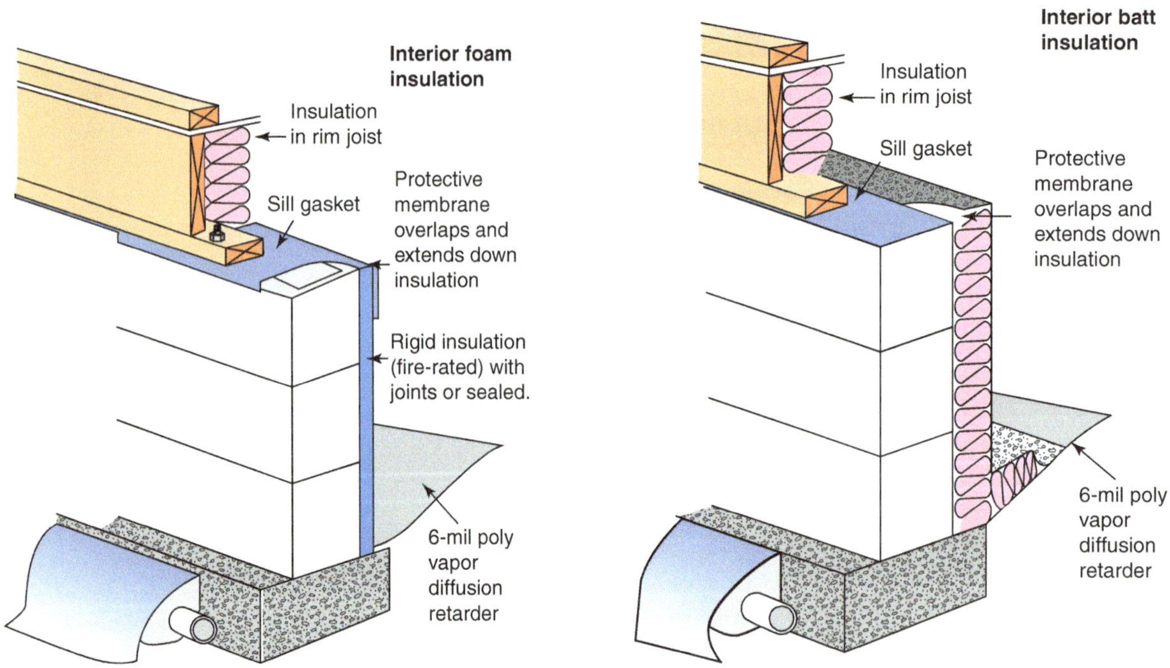

FIGURE 15-18 Interior insulated crawl space. *(Courtesy of EnergySavers.gov)*

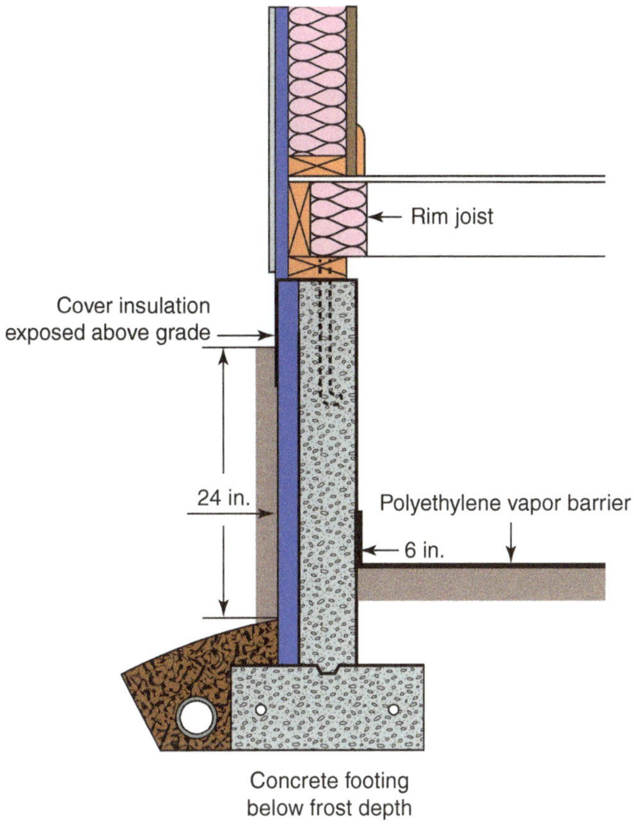

FIGURE 15-19 Exterior insulated crawl space.

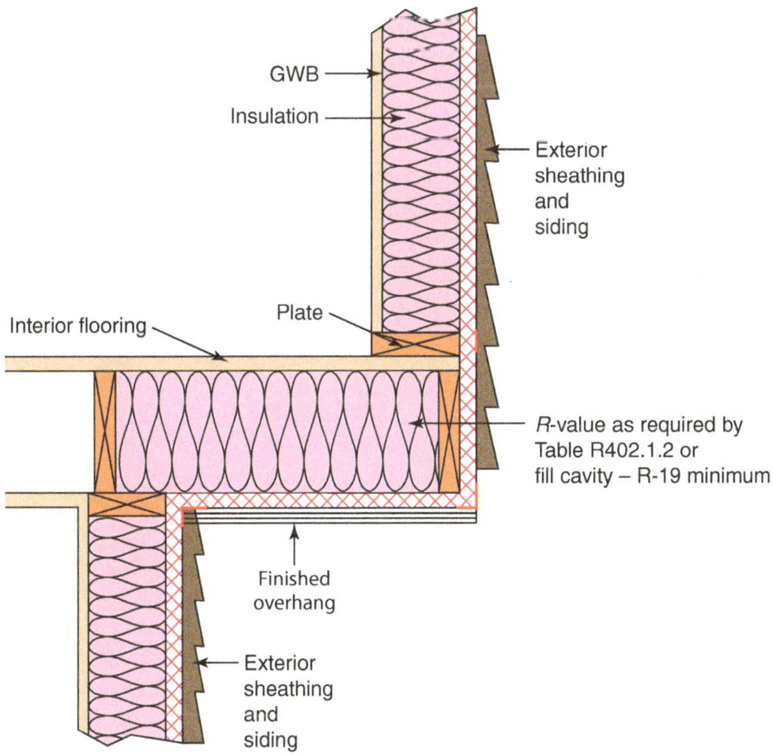

FIGURE 15-20 Cantilever floor insulation.

FLOORS

Insulation requirements for wood-frame floor joist construction are easy to find in the now-familiar *R*-value and *U*-value tables in Section R402. Table 402.1.2 includes an important footnote g that is an exception to the minimum insulation requirements. In climate zones where R-30 or R-38 insulation is required, R-19 is an acceptable minimum. The framing cavity must be filled and depending on the depth of the floor joists, the *R*-value may be greater than R-19. This footnote acknowledges that increasing joist size to accommodate R-30 or R-38 insulation adds considerable cost to construction and is not justified in energy savings over the useful life of the building.

Floors framed with steel joists are required to meet the equivalent *R*-values listed in Table R402.2.6. Notice that in all but one application, continuous insulation is required to comply with the steel-frame table for floors. The constructability of this assembly must be carefully considered during the design phase of plan development. Sufficient detail must be included in the plan documents to demonstrate compliance in the permit review and field inspection approvals.

In both wood- and steel-frame floor assemblies, the insulation must be installed to maintain permanent contact with the underside of the subfloor above (Figure 15-20). If the insulation sags or droops, cold air will move between the insulation and floor, minimizing the effectiveness of the insulation and creating uncomfortable cold spots in the floor (Figure 15-21). **[Ref. R402.2.8]**

The *R*-value, *U*-factor and installation requirements apply to all floors that are part of the thermal envelope and do not apply to floors with conditioned space both above and below the floor.

Slab-on-grade floors

Many buildings are designed with slab-on-grade floors. This technique saves the cost of excavating the site to accommodate a basement or crawl space. It is also possible to incorporate the slab on grade in portions of the design and crawl space or basement construction in the same building. Building site slope and topography usually dictate the decisions designers and builders make to maximize the building's integration with its surroundings.

Any portion of the conditioned space that incorporates a slab 12 inches or less below grade requires insulation. The insulation creates a thermal break between the interior conditioned space and earth adjacent to the concrete footing and slab. Heated slabs require additional R-5 insulation at the slab edges. [Ref. R402.2.10] The U.S. Department of Energy (DOE) estimates that slab edge insulation can reduce winter heating bills by 10 to 20 percent, and in new construction the annual energy savings exceeds the added cost of construction when factored into the mortgage cost. The investment in slab insulation pays off from the beginning.

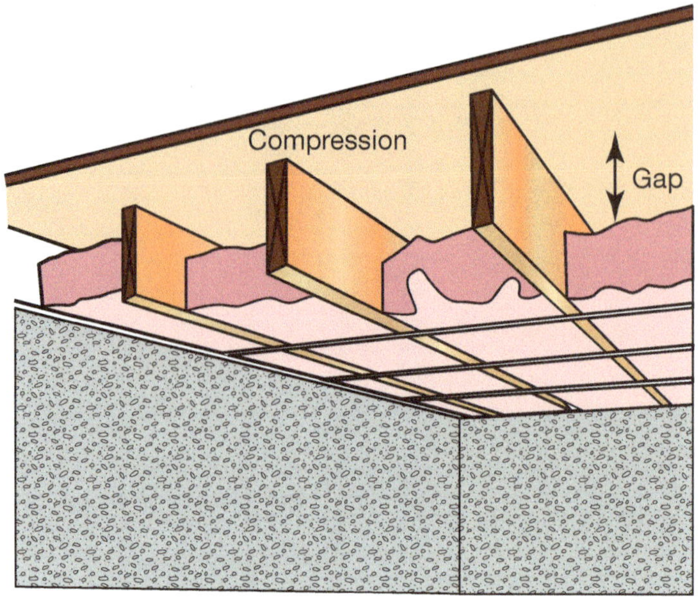

Two common flaws in floor insulation are gaps above the batt and compression of the batt in the cavity.

FIGURE 15-21 Sagging floor insulation.

Unheated slabs in Climate Zones 1, 2 and 3 do not require insulation. Notice that the *U*-factor Table R402.1.4 does not offer an alternative to the *R*-value and depth requirements in Table R402.1.2. Slab-on-grade design serving conditioned spaces must meet the minimums indicated in Table 402.1.2 (Figure 15-22). Footnote d is important, as it notes that an additional R-5—for a total of R-15 (10 + 5)—is required for all edges of heated slabs as heat is lost when the edge is exposed to outside air. The insulation must begin at the top of the slab. Heated slabs in Climate Zones 1, 2 and 3 do not require slab edge insulation, where the R-5 additional insulation alone is adequate. The insulation depth of 2 or 4 feet is determined by climate zone (Figure 15-23). [Ref. Table 402.1.2]

The only exception allowed for slab-on-grade insulation is in localities with a very heavy infestation of termites. Termite infestation probability maps are

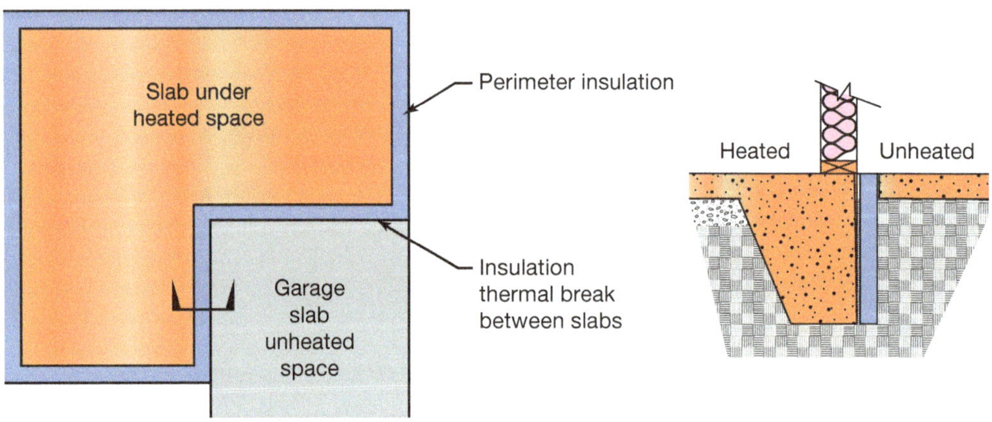

FIGURE 15-22 Mixed slab.

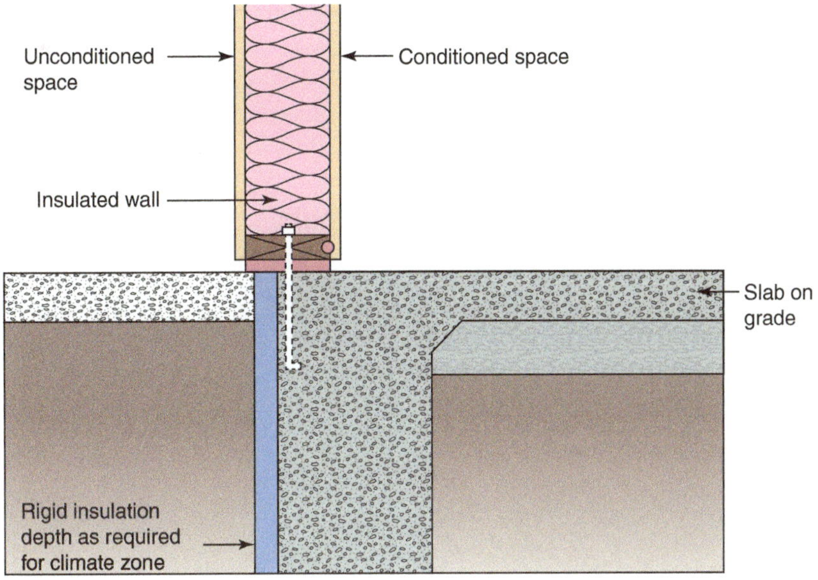

FIGURE 15-23 Insulation detail.

found in IRC Figure R301.2(7) and IBC Figure 2603.8. However, every building mapped in the "very heavy" areas is not automatically exempt from the insulation requirements. It is very important to check with the code official in the jurisdiction where the proposed building will be constructed.

SUNROOMS

Sunrooms are a special case in the energy code. These special-use rooms are defined in Section R202 as "A one-story structure attached to a dwelling with a glazing area in excess of 40 percent of the gross area of the structure's *exterior walls* and roof." The driving statement in the code requires all conditioned sunrooms to meet the insulation requirements. The key design elements for conditioned sunrooms are found in the exceptions for sunrooms with thermal isolation. A thermally isolated sunroom is physically separated from the dwelling and the heating and cooling is controlled by a separate zone thermostat or by separate equipment (Figure 15-24). Wall insulation requirements are the same, R-13, for all climate zones, while ceiling requirements vary. If the sunroom is not thermally isolated (Figure 15-25), the wall and ceiling insulation must meet requirements set by the *R*-value or *U*-factor tables. **[Ref. R402.2.13]**

The windows, doors and skylights in sunrooms must meet the minimum requirements for the applicable climate zone except when the attached space is not thermally isolated. The *U*-factor for fenestration is relaxed for thermally isolated sunrooms in Climate Zones 2 through 8.

FIGURE 15-24 Thermally isolated sunroom.

FIGURE 15-25 Nonthermally isolated sunroom.

Vertical fenestration are allowed to have a maximum *U*-factor of 0.45 and skylights a maximum of 0.70. **[Ref. R402.3.5]**

FENESTRATION

When used in architecture, fenestration refers to windows in a building. Fenestration has an energy code-specific definition that makes it an all-encompassing term for light-transmitting areas in a building's thermal envelope. The definition includes windows, skylights and opaque and glazed doors. Fenestration performance affects occupant comfort and building energy use because the insulation value of a window is much lower than that of the wall it is installed in.

Windows

Table R402.1.2 illustrates the relationship between window and skylight energy-saving performance characteristics and climate zones (Table 15-6). Notice that the *U*-factor requirements range from "not required" (NR) in semi-tropical climates to 0.32 in semi-arctic climates. The required insulating quality of windows and skylights becomes more restrictive as the climate zone's Heating Degree Days (HDD) increases. Conversely, the maximum SHGCs begin at 0.25 in the warmest climate zones, where the Cooling Degree Days (CDD) dominate the energy-use load and nearly every building is air conditioned. Solar heat gain can be a good thing in cooler climates and can be maximized with thoughtful and proper design. Table R402.1.2 reflects that wisdom in the five climate zones with no requirement for SHGC performance.

The *U*-factor and SHGC rating of windows, doors and skylights is determined according to the standards referenced in Section R303.1.3. Every product is tested and labeled with the *U*-factor, SHGC, visible transmittance (VT) and air leakage values. If the fenestration product is not tested and labeled, the default values in Tables R303.1.3(1), (2) and (3) must be used to determine compliance.

TABLE 15-6 Table R402.1.2: Insulation and fenestration requirements by component.

Climate zone	Fenestration *U*-factor[a]	Skylight *U*-factor	Glazed fenestration SHGC[a,b]
1	NR	0.75	0.25
2	0.40	0.65	0.25
3	0.35	0.55	0.25
4 except Marine	0.35	0.55	0.40
5 and Marine 4	0.32	0.55	NR
6	0.32	0.55	NR
7 and 8	0.32	0.55	NR

a. The fenestration *U*-factor column excludes skylights. The SHGC column applies to all glazed fenestration. Exception: In Climate Zones 1 - 3, skylights shall be permitted to be excluded from glazed fenestration SHGC requirementsrovided that SHGC for such skylights does not exceed 0.30.

b. There are no SHGC requirements in the Marine Zone.

You Should Know

The National Fenestration Rating Council is a nonprofit organization providing a labeling system for the energy performance of windows, doors and skylights. The ratings establish a uniform method to compare fenestration products and design benefits desirable for climate, location and building orientation.

The numbers in the boxes on the label use the terms defined in the energy code. [Ref. R202] Find these terms in the code book, and then compare the code's definitions to the following information from the U.S. Environmental Protection Agency (EPA) Energy Star publication:

- ***U*-Factor** measures the rate of heat transfer and tells you how well the window insulates. *U*-factor values generally range from 0.25 to 1.25 and are measured in Btu/h·ft²·°F. The lower the *U*-factor, the better the window insulates.
- **Solar Heat Gain Coefficient (SHGC)** measures the fraction of solar energy transmitted and tells you how well the product blocks heat caused by sunlight. SHGC is measured on a scale of 0 to 1; values typically range from 0.25 to 0.80. The lower the SHGC, the less solar heat the window transmits.
- **Visible Transmittance (VT)** measures the amount of light the window lets through. VT is measured on a scale of 0 to 1; values generally range from 0.20 to 0.80. The higher the VT, the more light you see.
- **Air Leakage (AL)** measures the rate at which air passes through joints in the window. AL is measured in cubic feet of air passing through one square foot of window area per minute. The lower the AL value, the less air leakage. Most industry standards and building codes require an AL of 0.3 cf·m/ft².
- **Condensation Resistance** measures how well the window resists water build-up. Condensation Resistance is scored on a scale from 0 to 100. The higher the Condensation Resistance factor, the less buildup the window allows. ●

Not every window or skylight has to comply with *U*-factor and SHGC maximum requirements when using the area-weighted-average *U*-factor approach. [Ref. R402.3.1] The concept of averaging glazing performance and the total square feet in the proposed design is straightforward and easily illustrated in a simple example:

Example:

These window types are included in the schedule for a single-family home proposed in Climate Zone 7. The *U*-factor is noted on the window schedule.

Window type 1	U-0.30	125 ft²
Window type 2	U-0.32	75 ft²
Window type 3	U-0.35	25 ft²

Answer: $$\frac{(125\text{ ft}^2 \times .30) + (75\text{ ft}^2 \times .32) + (25\text{ ft}^2 \times .35)}{225\text{ ft}^2} = \text{U-0.312 average}$$

U-0.32 is required in Climate Zone 7, and 0.312 is less than 0.32, so the proposed window selection complies using an area-weighted average calculation.

The area-weighted compliance approach to SHGC is similar to but not exactly like the *U*-factor method. Only fenestration that is more than 50 percent glazed is to be included in the calculation. Entry doors, for example, may be less than half glass, whereas patio doors may be mostly glass. This makes sense, as the heat gain potential is a direct function of the amount of glazing in the fenestration product. [Ref. R402.3.2]

Example:

These window types are included in the schedule for a single-family home proposed in Climate Zone 3. The SHGC value is noted on the window schedule.

Window type 1	SHGC-0.23	125 ft^2
Window type 2	SHGC-0.25	75 ft^2
Window type 3	SHGC-0.28	25 ft^2

Answer: $$\frac{(125\ ft^2 \times .23) + (75\ ft^2 \times .25) + (25\ ft^2 \times .28)}{225\ ft^2} = \text{SHGC } 0.242 \text{ average}$$

SGHC 0.25 is required in Climate Zone 3, and 0.242 is less than 0.25, so the proposed windows comply using an area-weighted average calculation.

The code does place limits on the performance of fenestration allowed in the area-weighted calculation method to demonstrate compliance. The mandatory provisions in Section R402.5 list maximum weighted average *U*-factors and SHGC values for vertical fenestration and skylights by climate zone. This code section eliminates the exclusive use of any fenestration product that exceeds these maximums in the two tradeoff methods to demonstrate compliance. Fenestration *U*-factors and SHGC values in designs using the simulated performance method are limited by the restrictions in this mandatory section. **[Ref. R402.5]**

Dynamic glazing can be permitted to satisfy the SHGC requirements of Table R402.1.2 provided that the ratio of maximum to minimum SHGC is greater than or equal to 2.4 and the dynamic glazing is automatically controlled to modulate the amount of solar heat gain into the space in multiple steps. Automatic control can be regulated by either electrical systems or environmental factors. This type of glazing should be considered separately from nondynamic glazing area-weighted averaging. **[Ref. R402.3.2]**

Dynamic glazing systems employ a variety of methods to vary the SHGC and VT of the fenestration product. Commonly encountered dynamic glazing assemblies make use of internal shading or electrochromic technology. Internal shading is self-explanatory – there is a shade, like those commonly mounted on the interior of fenestration, enclosed between glazing panes. The shades can employ a combination of variable louvers and retractability to modulate the properties of the fenestration product. Electrochromic dynamic glazing modulates the transmitted light through application of a direct current (DC) potential to the specially constructed layers on the glazing which causes nano-particles to rearrange. The new configuration has a lower VT and creates a shading effect. This is mainly used when shades and other methods that fully obscure outside visuals want to be avoided.

Every door and window is required to be labeled with the NFRC 100 test results. The field inspector will check each unit for compliance with the window and door schedule. If the product has not been tested, the values in the three default tables will determine the rating (Tables 15-7, 15-8, 15-9). It is important to note that these default values exceed the maximum average values allowed to be used in the area-weighted calculation.

TABLE 15-7 Table R303.1.3(1): Default glazed fenestration *U*-factor.

Frame type	Single pane	Double pane	Skylight	
			Single	Double
Metal	1.20	0.80	2.00	1. 30
Metal with thermal break	1.10	0.65	1.90	1.10
Nonmetal or metal clad	0.95	0.55	1.75	1.05
Glazed block		0.60		

TABLE 15-8 Table R303.1.3(2): Default door *U*-factors.

Door type	*U*-factor
Uninsulated metal	1.20
Insulated metal	0.60
Wood	0.50
Insulated, nonmetal edge, max 45 percent glazing, any glazing double pane	0.35

TABLE 15-9 Table R303.1.3(3): Default glazed fenestration SHGC and VT.

	Single glazed		Double glazed		Glazed block
	Clear	Tinted	Clear	Tinted	
SHGC	0.8	0.7	0.7	0.6	0.6
VT	0.6	0.3	0.6	0.3	0.6

Doors

The energy code offers two very practical exemptions to all of the fenestration requirements. The intent is to allow some design flexibility for designers and contractors to include decorative windows, skylights and glazed doors. Up to 15 square feet of glazing per dwelling is exempt from both the *U*-factor and SHGC code requirements, but there is a condition: The exemption can only be taken using the fenestration component requirements specified in Table R402.1.2. **[Ref. R402.3.3]** A similar exemption is available to exclude a side-hinged opaque door assembly no bigger than 24 square feet. Again, it is important to understand that the design does not qualify for these two exemptions if the *U*-factor alternative or Total UA alternative paths are used to demonstrate energy code compliance. **[Ref. R402.3.4]**

FIGURE 15-26 Window replacement. *(Photo courtesy of Ernie Decker/thinkstockphotos.com)*

Fenestration replacement

In work that requires removing a window sash, a skylight window frame or glazed door, the new fenestration components are required to comply with the *U*-factors and SHGC values for new construction. [Ref. 503.1.1.1] This is an important distinction when considering Item 1 in Section R504.2 that allows any type of glass pane to be installed in an existing sash. When the sash and/or frame are replaced, the new units must meet the applicable requirements in Table R402.4.1.1 (Figure 15-26). When only a window pane is replaced, the glazing is not required to comply (Figure 15-27).

FIGURE 15-27 Window pane replacement glazing is not regulated by the IECC.

Infiltration

Air leakage in the the building envelope is called infiltration even though it is a combination of infiltration and exfiltration. Air moves into and out of the the conditioned space when the pressure inside the thermal envelope is less than or greater than the air pressure outside. Exfiltration describes air movement through the building thermal envelope to the outside. Differentials in pressure and temperature cause air movement through the building thermal envelope if it is not properly sealed. This allows conditioned air to escape and creates a particularly inefficient building energy use. Infiltration and exfiltration occur at the same time while the building works to balance itself, with air finding cracks and holes in the *air barrier*.

Many building designs rely on natural ventilation, meaning an open or closed window, to provide critical building pressure control. Other design strategies incorporate a balanced controlled ventilation system. Heat recovery ventilation (HRV) and energy recovery ventilation (ERV) supply fresh air to the building interior and "recover" energy from the tempered exhaust air. HRV and ERV systems prove to be very efficient and effective in buildings with very little air leakage.

The 2018 IECC recognizes that air leakage control is absolutely critical in the "construction of buildings for the effective use and conservation of energy over the useful life of each building." To achieve the stated intent of the energy code, certain provisions, including mandatory testing, are requirements. [Ref. R402.4]

THE AIR BARRIER AND BUILDING THERMAL ENVELOPE

The specific requirement for the continuous air barrier is that it must be installed in the building thermal envelope. An easy test to check for compliance is to trace the air barrier location in the building section details on the construction documents. If this can be done without lifting the pencil off the paper, the air barrier is continuous. Table R402.4.1.1 lists 16 building envelope elements that require special attention to detail.

FIGURE 15-28 Field mock-ups sometimes avoid misunderstandings.

FIGURE 15-29 Before proper air sealing of penetration.

FIGURE 15-30 After proper air sealing of penetration.

Construction document notes and callouts will address how the building construction complies with these requirements. Table 15-10 highlights items to be detailed on the construction documents. All penetrations must be properly sealed.

Caulk and sealant materials and methods in air-barrier construction must allow for thermal and moisture expansion and contraction. Joints between different materials have to shrink and stretch as the air-barrier components go through daily heat-up and cool-down cycles. Any material used must be installed per the manufacturer's installation instructions. The instructions must be on site and available for review during the field inspection process (Figure 15-28). **[Ref. R402.4.1]**

Fireplaces and recessed luminaires (light fixtures) merit special consideration in air-barrier detailing and inspection. New wood-burning fireplaces must have tight-fitting flue dampers and outdoor supplied combustion air. **[Ref. R402.4.2]** The damper requirement keeps conditioned air from flowing out the chimney, and the outdoor supplied combustion air requirement ensures the fireplace pulls in unconditioned outside air to burn instead of pulling conditioned air from inside the home. Fireplace construction requirements are also found in IRC Chapter 10 and IBC Chapter 21.

Recessed light fixtures installed in the thermal envelope must be air sealed. A gasket or caulk between the recessed light housing and the interior wall or ceiling is required. Because the lights are penetrating the thermal envelope, the units must be IC-rated (insulation contact), tested and labeled for air leakage. A sealed box over the can light fixture with proper detailing also complies. The box must be sealed to the attic floor, the wire penetrations into the box must be sealed, and the entire assembly must be installed with a minimum 3-inch clearance to adjacent insulation (Figures 15-29 and 15-30). **[Ref. R402.4.5]**

In Climate Zones 3 through 8, rooms containing fuel-burning appliances must be sealed and insulated from the conditioned portions of the building in accordance with the building thermal envelope provisions of the code. **[Ref. R402.4.4]**

TABLE 15-10 Table R402.4.1.1 (2018 IECC): Air barrier and insulation installation.

Component	Air barrier criteria	Insulation installation criteria
General requirements	A continuous air barrier shall be installed in the building envelope The exterior thermal envelope contains a continuous air barrier. Breaks or joints in the air barrier shall be sealed.	Air-permeable insulation shall not be used as a sealing material
Ceiling/attic	The air barrier in any dropped ceiling/soffit shall be aligned with the insulation and any gaps in the air barrier shall be sealed. Access openings, drop down stairs or knee wall doors to unconditioned attic spaces shall be sealed	The insulation in any dropped ceiling/soffit shall be aligned with the air barrier.
Walls	The junction of the foundation and sill plate shall be sealed. The junction of the top plate and the top of exterior walls shall be sealed. Knee walls shall be sealed.	Cavities within corners and headers of frame walls shall be insulated by completely filling the cavity with a material having a thermal resistance, *R*-value, of not less than R-3 per inch. Exterior thermal envelope insulation for framed walls shall be installed in substantial contact and continuous alignment with the air barrier
Windows, skylights and doors	The space between framing and skylights, and the jambs of windows and doors, shall be sealed.	—
Rim joists	Rim joists shall include the air barrier.	Rim joists shall be insulated.
Floors (including cantilevered floors and floors above garages)	The air barrier shall be installed at any exposed edge of insulation.	Floor framing cavity insulation shall be installed to maintain permanent contact with the underside of subfloor decking. Alternatively, floor framing cavity insulation shall be in contact with the top side of sheathing, or continuous insulation installed on the underside of floor framing; and shall extend from the bottom to the top of all perimeter floor framing members.
Crawl space walls	Exposed earth in unvented crawl spaces shall be covered with a Class I vapor retarder with overlapping joints taped.	Crawl space insulation, where provided instead of floor insulation, shall be permanently attached to the walls.
Shafts, penetrations	Duct shafts, utility penetrations, and flue shafts opening to exterior or unconditioned space shall be sealed.	—
Narrow cavities	—	Batts to be installed in narrow cavities shall be cut to fit, or narrow cavities shall be filled with insulation that on installation readily conforms to the available cavity space.
Garage separation	Air sealing shall be provided between the garage and conditioned spaces.	—
Recessed lighting	Recessed light fixtures installed in the building thermal envelope shall be sealed to the finished surface.	Recessed light fixtures installed in the building thermal envelope shall be air tight and IC rated.
Plumbing and wiring	—	In exterior walls, batt insulation shall be cut neatly to fit around wiring and plumbing, or insulation, that on installation readily conforms to available space, shall extend behind piping and wiring.
Shower/tub on exterior wall	The air barrier installed at exterior walls adjacent to showers and tubs shall separate the wall from the shower or tub.	Exterior walls adjacent to showers and tubs shall be insulated.
Electrical/phone box on exterior walls	The air barrier shall be installed behind electrical and communication boxes. Alternatively, air-sealed boxes shall be installed.	—
HVAC register boots	HVAC supply and return register boots that penetrate building thermal envelope shall be sealed to the subfloor, wall covering or ceiling penetrated by the boot.	—
Concealed sprinklers	Where required to be sealed, concealed fire sprinklers shall only be sealed in a manner that is recommended by the manufacturer. Caulking or other adhesive sealants shall not be used to fill voids between fire sprinkler cover plates and walls or ceilings.	—

a. Inspection of log walls shall be in accordance with the provisions of ICC-400.

BLOWER DOOR TEST

A blower door test is mandatory to verify that the requirements for maximum air leakage are met. The calibrated blower door assembly is elegant and simple; an easily portable instrument fits into an exterior door frame. A powerful variable-speed fan pulls air out of the building or dwelling being tested, and outside air flows inside through unsealed cracks and holes (Figure 15-31). The speed of the fan is automatically adjusted to maintain a specified pressure differential, usually 0.2 inch w.g., between the inside of the dwelling and outside. The higher the fan speed, the less effective the air barrier. Specific test criteria must be followed to maintain comparable and consistent measurement results. Preparation and criteria for the test and required performance are addressed in R402.4.1.2.

FIGURE 15-31 Blower door test being performed. *(Photo courtesy of Retrotec)*

The blower door test must confirm that the total air leakage is less than 5 air changes per hour (ACH) in Climate Zones 1 and 2 and not more than 3 ACH in Climate Zones 3 through 8. The test results must be submitted to the code official in a written, signed report. **[Ref. R402.4.1.2]**

You Should Know

Unlike the commercial sections of the IECC, in the residential section there is no option for testing *or* visual inspection. The requirement is now testing *and* visual inspection. The code official is authorized to require an approved third party to inspect and verify. ●

You Should Know

Blaise Pascal was a hardworking French philosopher, mathematician and physicist in the mid-seventeenth century. In his short 39-year life, he developed a mathematical probability theory used to make gambling and economics more or less predictable, invented the first mechanical calculator, and may have worn one of the first wristwatches.[1] Pascal is obviously the namesake for the pascal (Pa) unit of measurement, a unit of atmospheric pressure equal to 1 newton per square meter, in the International System of Units (SI). Standard atmospheric pressure is 101,325 Pa.

Blower door tests are calibrated to a 50-Pa pressure difference between the inside and outside of the building air barrier. Fifty Pa of pressure is equal to .00725 psi, which is not very much. One pascal of pressure is about the weight of one postage stamp laying flat over its roughly 1 in^2 in area. ●

[1]William J. Lynott, *Watch Development,* February 2012.

CHAPTER 16

Building Systems

© A S Klein Engineering, PLLC

Mechanical ventilation is a mandatory requirement in the 2018 *International Energy Conservation Code®* (IECC®). The requirement is referenced in Section R303.4 of the *International Residential Code®* (IRC®) and Section 401.2 of the *International Mechanical Code®* (IMC®). Both codes state that mechanical ventilation is required when the air infiltration rate in the dwelling is less than 5 air changes per hour (ACH). These code provisions align requirements regarding mechanical ventilation. Every mechanical system must be designed with controls, delivery mechanisms and properly sized equipment. Specific provisions that support energy savings in heating, ventilating and air conditioning (HVAC) and hot water systems are also included.

PROGRAMMABLE THERMOSTAT

Energy is conserved in dwellings by adjusting the demand for heating and cooling correctly. All forced-air furnace heating and cooling systems must have at least one programmable thermostat. It must be capable of daily time and temperature settings. Settings for zones range from 55°F to 85°F. This required feature may be used as a "vacation" from typical heating and cooling requirements and can compound energy savings when the dwelling is unoccupied. The programmable thermostat must be specified to limit the heating temperature to no higher than 70°F and the cooling temperature to no lower than 78°F, and must be referenced and called out in the plan documentation. This is the initial setting that is verified and approved at the final building inspection.

DUCTS

Ducts are the delivery passageways for the ventilation system that move fresh air into and stale air out of the dwelling. The effectiveness of the ductwork is critical to efficient energy use in the building and providing comfort for the people living in the building. Ducts must transfer conditioned air under positive pressure from the air-handling unit to rooms around the building and take stale air out under negative pressure for recirculation or exhaust. Provisions specific to ducts and air handlers improve energy efficiency in the design and installation of these systems.

Supply ducts must be insulated unless installed completely inside the building thermal envelope. **[Ref. 403.3]** R-8 insulation is required to protect the conditioned air moving through supply ducts greater, less than or equal to 3 inches in diameter in unconditioned attics. This helps prevent condensation (sweating) in warmer months on air-conditioner ducts and keep heated air warmer in the cooler months. Insulated air-tight supply and return ducts may be laid on the ceiling directly below the attic and covered with blown-in insulation. A R-8 minimum insulated duct is required. The minimum ceiling insulation values of R-30, R-38 or R-49 in Table R402.1.2 must be met. Best practice is to install as shown in Figure 16-1.

A more robust provision to minimize condensation in Climate Zones 1A, 2A and 3A (hot-humid and mixed-humid climates) allows R-13 ducts to be buried in blown-in attic insulation. Although not specifically mentioned in the code provision, the best practice is also to insulate the duct boots to the minimum required insulation value and install

Code Essentials

Multiple system designs are possible for an air distribution system to deliver code-compliant energy efficiency and provide a comfortable thermal environment. Duct sizing is required by IMC Section 603.2 and references ACCA *Manual D* as an approved method. Duct sizing methods consider building floor plan, room sizes, number of levels, the size of windows in each room, and forced air furnace location. ACCA *Manual D* establishes design parameters and calculation methods to provide an air-handling system for comfortable airflow to each room.

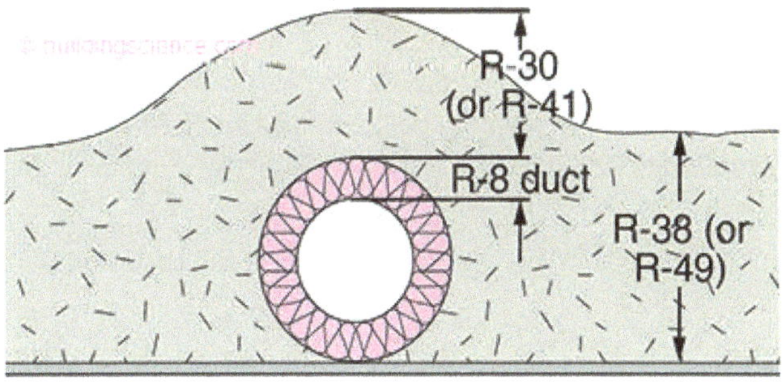

FIGURE 16-1 R8 Blown-in duct.

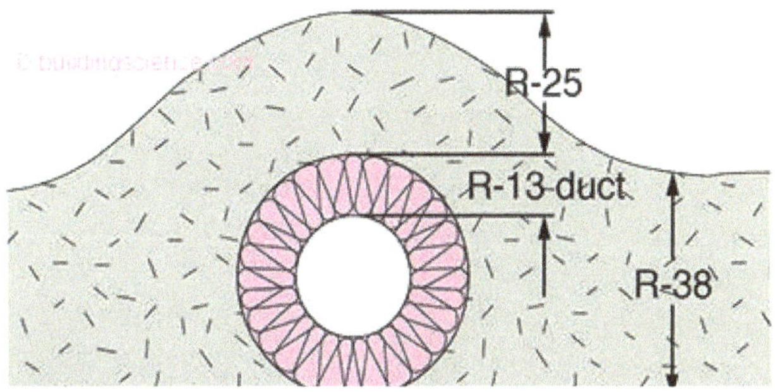

FIGURE 16-2 R-13 Blown-in duct.

FIGURE 16-3 Duct-seal mastic.

as shown in Figure 16-2. R-13 supply ducts must meet the minimum vapor retarder provisions of the IRC or IMC and the minimum ceiling insulation values of R-30 in Climate Zone 1 or R-38 in Climate Zones 2 and 3 to comply with this provision. **[Ref. 403.3.6]**

Ductwork greater than or equal to 3 inches in diameter located elsewhere that is not completely inside the building thermal envelope must be installed with at least R-6 insulation. Ductwork less than 3 inches in diameter has a reduced insulation requirement of R-6 in attics and R-4.2 elsewhere.

Many HVAC energy experts estimate[1] that about 20 percent of conditioned air intended for distribution in the dwelling unit does not make it to the room or space due to leaks, holes and poorly constructed ductwork systems. These losses result in higher energy bills and a lower level of occupant comfort. Ducts, air handlers and filter boxes must be sealed and inspected during construction (Figure 16-1). The IMC and IRC give direction as to compliant materials, techniques and standards regarding duct construction. These requirements apply to all ducts in all construction. Installation of flexible ducts is covered by provisions in IMC Section 603. All materials must be installed according to the manufacturer's specifications (see Figure 16-7). The IECC requires that duct system air-tightness be tested and verified at either the mechanical rough-in or final inspection. Ducts and air handlers located entirely within the building thermal envelope do not have to be duct-blaster tested.

Designers, builders, plans examiners and building inspectors must be familiar with the requirements and specific exceptions applicable to mechanical system components intentionally located entirely within the building thermal envelope. Well-designed and properly sealed duct systems save energy and can make the dwelling more comfortable.

[1]www.energystar.gov (see Home Improvement—Duct Sealing).

EXAMPLE

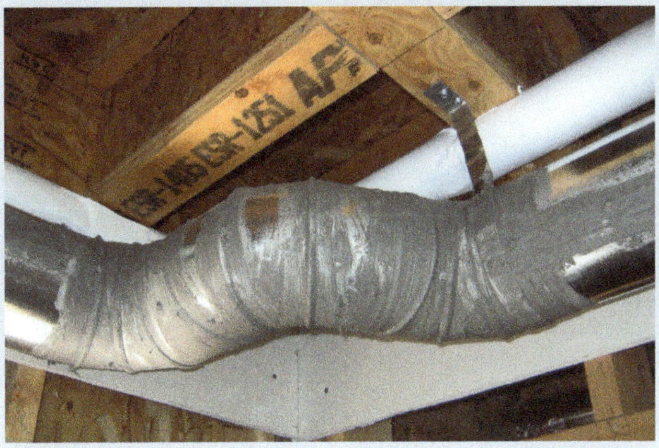

FIGURE 16-4 Twisted duct.

The duct construction in Figure 16-2 shows three Type 8AA elbow fittings used to achieve one 90-degree turn. How does this change the equivalent length in the duct flow calculation (Figure 16-3)?

Elbows and offsets			
Picture	ID	Equivalent length	Fitting description
	8A6	20	Smooth elbow, R/D = 0.75
	8A9	15	Smooth elbow, R/D = 1.0
	8AE	10	Smooth elbow, R/D = 1.5
	8AA	20	4 or 5 piece elbow, R/D = 1.0
	8AF	15	4 or 5 piece elbow, R/D = 1.5
	8A8	35	3 piece elbow, R/D = 0.75
	8AB	25	3 piece elbow, R/D = 1.0
	8AH	20	3 piece elbow, R/D = 1.5

FIGURE 16-5 *Manual D* duct sizing. *(Photo courtesy of ACCA)*

Answer: Each 8AA elbow adds 20 feet in equivalent duct length. The two additional elbows add another 40 feet. This will affect the static duct pressure and the amount of fan pressure needed in the duct to deliver the cubic feet per minute (CFM) in the room as required by ACCA *Manual D.*

HOT WATER SYSTEMS

Most mechanical system pipes designed to carry hot water are required to be insulated with a minimum of R-3 pipe insulation (Table 16-1), installed and protected against damage. Equipment maintenance, sunlight and moisture all pose a threat to the expected performance of pipe insulation over the useful life of the building. **[Ref. R403.5.3]**

FIGURE 16-6 Insulating hot water pipes reduces heat loss.

Insulating hot water pipes reduces heat loss (Figure 16-6). This concept is also a water-conserving measure because warmer water is delivered to the fixture without the cooler water running down the drain.

Heated water circulation systems are required to be provided with a circulation pump; gravity and thermosyphon circulation systems are prohibited. Controls for circulating hot water system pumps and electric heat-trace systems must be based on the demand for hot water in the occupancy. **[Ref. R403.5.1.1]** Demand recirculation system pump controls have their own control requirements, including responding to plumbing fixture use and limiting the temperature of water entering cold water piping to 104°F. **[Ref. R403.5.2]**

Automatic controls, temperature sensors and pumps are required to be accessible, and manual controls must be readily accessible.

You Should Know

A water distribution system having one or more recirculation pumps that pump water from a heated water supply pipe back to the heated water source through a cold water supply pipe shall be a demand recirculation water system. ●

TABLE 16-1 Required hot water pipe insulation.

1.	Piping $^3/_4$ inch and larger in nominal diameter
2.	Piping serving more than one dwelling unit
3.	Piping located outside the conditioned space
4.	Piping from the water heater to a distribution manifold
5.	Piping located under a floor slab
6.	Buried-in piping
7.	Supply and return piping in recirculation systems other than demand recirculation systems

FIGURE 16-7 Improperly installed flexible duct.

MECHANICAL VENTILATION

Mechanical ventilation systems are covered by requirements in the 2018 IECC. The designs are either exhaust-only, supply-only or balanced air-exchange systems (Table 16-2). System details regarding sources of supply air, locations of exhaust air and ventilation rates are provided in IRC M1505.4 and IMC Section 403. The IMC states that "the amount of supply air shall be approximately equal to the amount of return and exhaust air," and that "the system shall not be prohibited from producing negative or positive pressure." The energy code does not include requirements for system design.

Heat and energy recovery ventilation systems (HRVs and ERVs, respectively) are energy efficient and cost effective in extreme heating and cooling climates (Figure 16-5). Heat loss is minimized by using conditioned exhaust air to warm or cool fresh incoming air. The ERV heat exchanger transfers water vapor and heat energy. An HRV only transfers the heat energy. ERVs are most effective in warm, humid climates. HRVs (Figure 16-6) are most effective in cold, dry climates. Simple systems are easily incorporated in new building design.

Outdoor air intakes and exhausts must have automatic or gravity-operated dampers that close when the mechanical system is not in operation. The louvered vent terminations on clothes dryer and bathroom exhaust ducts are common examples of gravity-operated dampers. These comply with the provision.

Heating and cooling equipment sizing is important to integrating building systems in new designs. Properly sealed and insulated building envelopes require more consideration than "that's the way we always did it." Provisions mandated by the energy code for thermal- and air-leakage improvements allow for smaller HVAC equipment. Equipment that is correctly sized operates efficiently and saves energy costs and maintenance expense. *Manual J* of the Air Conditioning Contractors of America (ACCA) provides an accepted method for calculating heating and cooling loads. Check with the local code official for other approved sizing methods used in the jurisdiction.

Efficient fan motors save electric energy. "Efficacy" means "the power to produce an effect"; in this context, it refers to how many watts of electricity it takes to move a cubic foot of air per minute (CFM). Table R403.6.1 (Table 16-3) lists fan locations and minimum efficacy requirements. Information demonstrating compliance is required to be included in notes, mechanical plans or equipment schedules. Bathroom fans rated at 90 CFM or greater are required to be more efficient than smaller fans.

You Should Know

Building design offers several options for mechanical ventilation systems. Considerations for each system are summarized in Table 16-2. ●

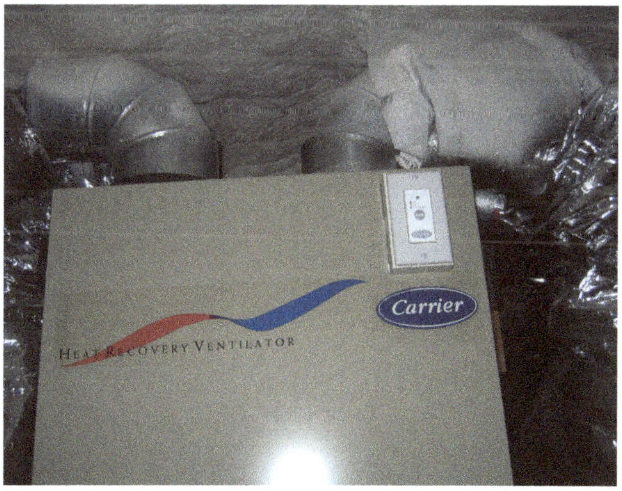

FIGURE 16-8 An HRV installed.

TABLE 16-2 Comparison of mechanical ventilation systems.

Ventilation type	Pros	Cons
Exhaust only Air is exhausted from the house with a fan	• Easy to install • Simple method for spot ventilation • Inexpensive	• Negative pressure may cause backdrafting • Makeup air is from random sources • Removes heated or cooled air
Supply only Air is supplied into the house with a fan	• Does not interfere with combustion appliances • Positive pressures inhibit pollutants from entering • Delivers to important locations	• Does not remove indoor air pollutants at their source • Brings in hot or cold air or moisture from the outside • Air circulation can feel drafty • Furnace fan runs more often unless fan has an ECM (variable-speed motor)
Balanced air exchange system Heat and energy recovery ventilators supply and exhaust air	• No combustion impact • No induced infiltration/exfiltration • Can be regulated to optimize performance • Provides equal supply and exhaust air • Recovers up to 80 percent of the energy in air exchanged	• More complicated design considerations • Overventilation unless the building is tight • Cost

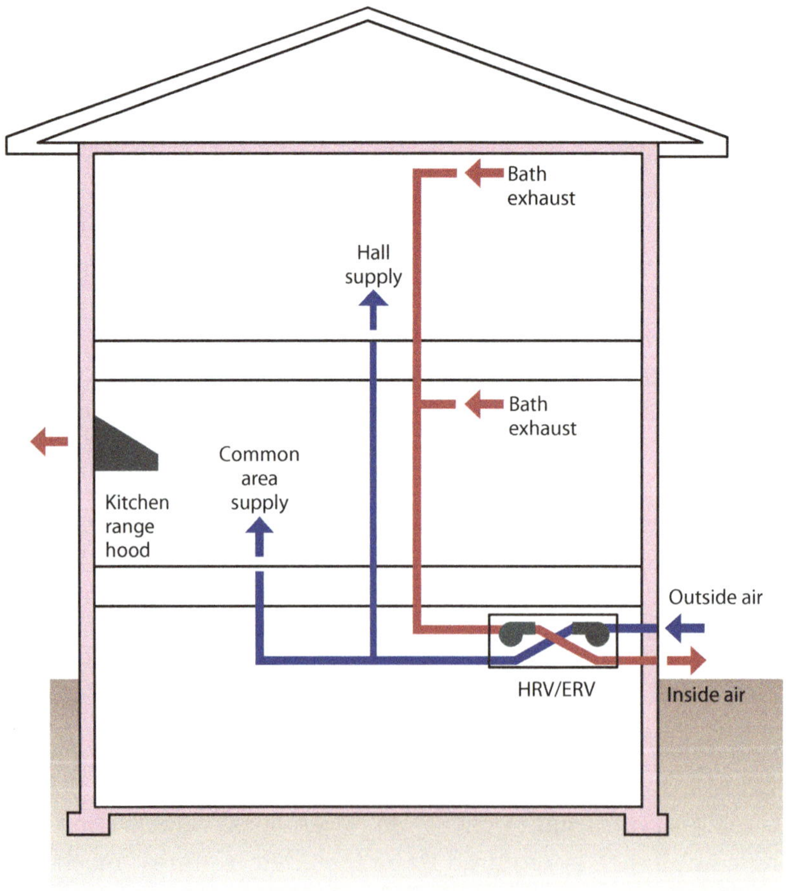

FIGURE 16-9 Simple ERV/HRV schematic layout. *(Courtesy of Building Science Corporation)*

SNOW MELT

Snow and ice accumulation on walkways, stairs, driveways and parking areas may create unsafe conditions around dwellings in cold climate zones (Figure 16-10). Operational controls for snow and ice systems are mandatory. Automatic controls include surface temperature sensors coupled with a moisture sensor (Figure 16-11). When the temperature is above 50°F at the surface sensors and it is not raining or snowing, the automatic control turns the system off. A manual or automatic control is also required to turn the system off when the outside temperature is above 40°F. The energy code does not reference standards or limit the size of snow melt installations. **[Ref. R403.9]**

FIGURE 16-10 Snow melt at work.

FIGURE 16-11 Snow melt control sensor.

POOLS AND SPAS

Energy-saving provisions for residential pool and spa pumps and heaters are addressed in the IECC. Vapor-retardant covers are required to limit evaporation, but not covers specifically installed to maintain the heated water temperature. This requirement is waived if more than 70 percent of the annual energy required to heat the pool is recovered on site.

Pool heaters must have an easy-to-reach on/off switch mounted on the outside of the heater or within 3 feet. Filter pumps and circulation motors are required to have control systems that automatically turn these devices on and off according to a preset schedule. Exceptions to the time switch provisions allow for public health requirements and pumps that operate pool heating systems using solar or waste heat recovery. **[Ref. R403.10]**

Portable spas must meet the requirements of the Association of Pool and Spa Professionals' (APSP) Standard APSP-14. **[Ref. R403.11]** Pools and permanent spas in attached and detached single-family dwellings are required to be in accordance with APSP-15. **[Ref. R403.12]**

TABLE 16-3 Table R403.6.1: Mechanical ventilation system fan efficacy.

Fan location	Airflow rate minimum (cfm)	Minimum efficacy (cfm/watt)	Airflow rate maximum (cfm)
Range hoods	Any	2.8 cfm/watt	Any
In-line fan	Any	2.8 cfm/watt	Any
Bathroom, utility room	10	1.4 cfm/watt	<90
Bathroom, utility room	90	2.8 cfm/watt	Any

For SI: 1 cfm = 28.3 L/min.

CHAPTER

17 Electrical Power and Lighting

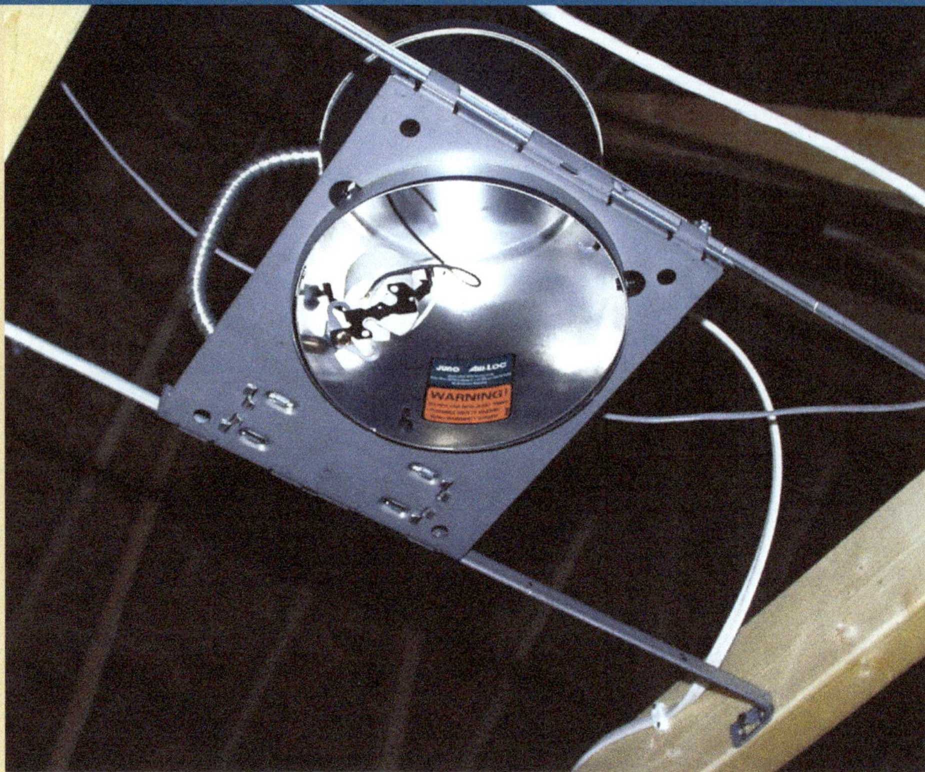

Lighting uses about 10 percent of residential electrical energy. A mandatory provision requires that at least 90 percent of installed light fixtures in a dwelling have high-efficacy lamps. [Ref. R404.1] The term "*high-efficacy*" regarding lighting fulfills the same energy-saving intent that the high-efficacy fan motor requirements serve in moving air. Ventilation and lighting, both necessary functions in a comfortable living environment, offer great opportunities for energy savings.

You Should Know

The 2014 Nobel Prize in physics was awarded to Isamu Akasaki, Hiroshi Amano, and Shuji Nakamura for their development of bright, blue LEDs in 1994. Green and red LEDs had previously been developed but to make white light lamps the blue is an integral part. It had taken almost 20 years from the creation of the first blue LED to construct one that shone bright enough for practical use. The official statement for why they received the prize is "for the invention of efficient blue light-emitting diodes which has enabled bright and energy-saving white light sources."

"The 2014 Nobel Prize in Physics - Press Release". Nobelprize.org. Nobel Media AB 2014. Web. 19 Oct 2017. ●

LIGHTING

The IECC defines high-efficacy lamps in lumens (light emitted) per watts used to produce the light. Lamps that produce 60 lumens/W for lamps over 40 W, 50 lumens/W for lamps over 15 W to 40 W and 40 lumens/W for lamps 15 W or less meet the provision (Table 17-1). **[Ref. 404.1]** A lamp is just the lightbulb itself. A chandelier is one fixture but may have many lamps. Lightbulb packages include the lumen information needed to determine compliance.

TABLE 17-1 High-efficacy lamps.

Lamp	Efficiency
≤ 15 W	40 lumens/W
>15 W–40 W	50 lumens/W
>40 W	60 lumens/W

TABLE 17-2 DoE lamps eff.

Comparisons between Traditional Incandescents, Halogen Incandescents, CFLs, and LEDs

	60W Traditional Incandescent	43W Energy-Saving Incandescent	15W CFL 60W Traditional	15W CFL 43W Halogen	12W LED 60W Traditional	12W LED 43W Halogen
Energy $ Saved (%)	-	-25%	-75%	-65%	-75%-80%	-72%
Annual Energy Cost*	$4.80	$3.50	$1.20		$1.00	
Bulb Life	1,000 hours	1,000 to 3,000 hours	10,000 hours		25,000 hours	

Based on 2 hrs/day of usage, an electricity rate of 11 cents per kilowatt-hour, shown in U.S. dollars.

"How Energy-Efficient Light Bulbs Compare with Traditional Incandescents." Department of Energy, energy.gov/energysaver/how-energy-efficient-light-bulbs compare-traditional-incandescents.

Example:

The information on a package of 40-watt lightbulbs indicates that the lamp produces 445 lumens. Is this a high-efficacy lamp?

Answer: 445 lumens/40 watts = 11 lumens per watt
11 lumens/watt is less than 50 lumens/watt, so this is not a high-efficacy lamp.

The newest innovation in home lighting technology is the Light-Emitting Diode (LED), which are 72 to 80 percent more energy efficient than their dated incandescent counterparts and have an improved lifespan over fluorescent bulbs. The Department of Energy created a comparison table for average lamps that produce similar light levels. To generate light, LEDs use semiconductor technology, the same principles, and materials that allow our computers and phones to be so small and portable. When electrons jump down from high to low energy states, they emit the difference in energy as light. In a fundamental sense, it is a controlled spark between wires. Instead of wasting energy heating a filament or gas, the conversion from electricity to light is almost a direct exchange.

Longer life expectancy is a product of LEDs not requiring conventional glass bulbs or tubes to contain a vacuum (incandescent) or gas (fluorescent), which are prone to leaks that make the bulbs inefficient or inoperable. LEDs do not require any gas to operate; their housings are typically an epoxy or other translucent materials. Because of their construction, LEDs are naturally directional, which means they do not require reflective surfaces that mitigate efficient lighting in applications such as overhead lamps and task lighting. To create a more desirable lamp for residential use, LEDs commonly are placed in traditional-looking bulbs that diffuse light and fill a room better.

Since LEDs emit light at specific wavelengths, it requires three different diodes to create white light. Fortunately, a single diode can be roughly the size of a grain of sand, and this is not a prohibitive factor.

Most fluorescent lamps meet the high-efficacy requirement; most incandescent bulbs do not (Figure 17-1). Light from fluorescent lamps "feels" different and is often described as "cool-white" or "warm-white" light. For most residential applications, it is usually appropriate to specify warmer lamp colors (CCT = 2700–3000 K), as they give a warmer feel.

Color temperature is the standard method to describe the visible range of light, and looks like a rainbow (Figure 17-2). The color of light emitted is measured and expressed in Kelvin (K) units, and this infor-

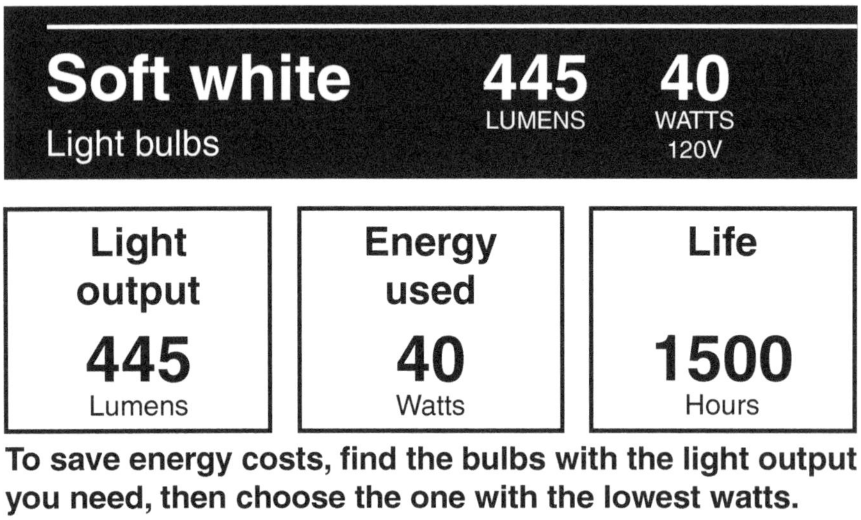

FIGURE 17-1 Incandescent bulbs do not meet high-efficacy requirements.

mation can be found on most lightbulb packages. The information for the 15 W compact fluorescent lamp (CFL) in Figure 17-3 indicates that the light measures 2700 K and closely resembles a typical household lightbulb.

At least 90 percent of the lamps in permanently installed lighting fixtures are required to be high-efficacy. Low-voltage lighting systems generally operate at 30 volts or less. [Ref. R404.1]

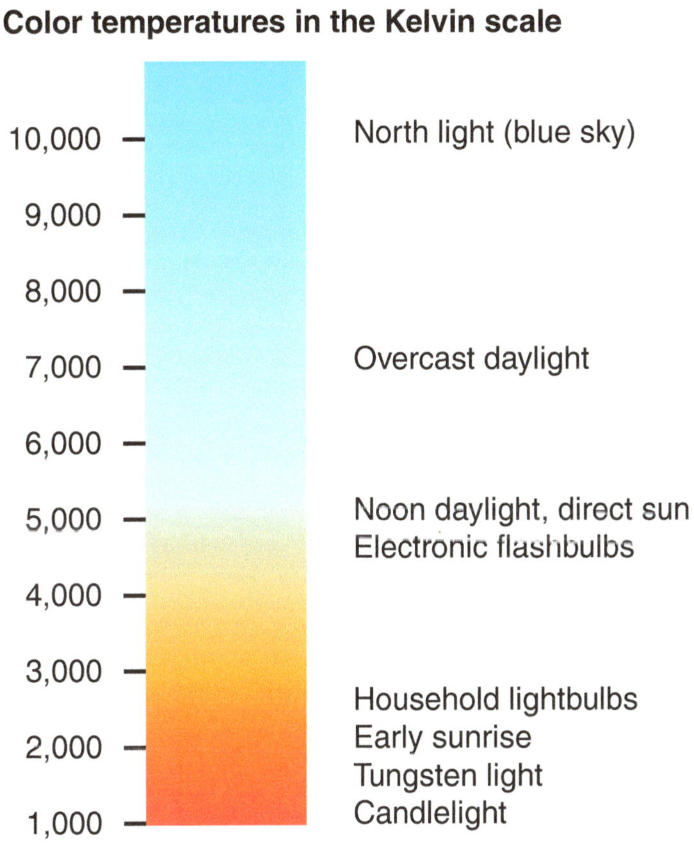

FIGURE 17-2 The color of light emitted is measured in Kelvin (K) units.

Compact fluorescent lamp

Mini spiral	950 lumens	15 watts	10,000 hours
Incand. bulb	800 lumens	60 watts	1,000 hours

FE – IIS – 15W
15W E26 2700K
120V 60Hz 275mA

FIGURE 17-3 Compact fluorescent lightbulb.

You Should Know

There are two ways to calculate lighting efficiency compliance: lamp count and fixture count. The 75-percent calculation is made for all electrical lighting fixtures that are not covered by the low-voltage exception. ●

FIGURE 17-4 Typical LED bulbs. (Courtesy of the Department of Energy)

Glossary

A

above-grade wall – In the energy code for residential buildings, when more than 50 percent of a wall is above grade, it is considered an above-grade wall. The measurement is the percentage of the portion of the building foundation or other exterior wall enclosing conditioned space above or below grade. For commercial prescriptive requirements, see also Section C402.2.2

accessible – Equipment located such that it is not in a locked room and is easy to approach. Readily accessible means the equipment can be maintained and inspected without crawling over other equipment or using a portable ladder. (Not related to accessible as in accessibility for the disabled.)

addition – In the energy code, an increase in the volume of conditioned space. This may be in floor area or height.

adhesive – A substance that produces a steady and firm attachment or adhesion between two surfaces.

air barrier – Material(s) that provide(s) a barrier to air leakage through the building envelope.

air changes per hour (ACH) – In a room or space, the complete replacement of air volume in a 1-hour period.

air conditioning – The mechanical process of treating air by controlling its temperature, humidity and distribution.

air exfiltration – Uncontrolled exit of air from a room or enclosed space to outdoors.

air infiltration – Uncontrolled entry of outside air into a room or an enclosed space.

American Recovery and Reinvestment Act (ARRA) – On February 13, 2009, in response to the U.S. economic crisis, Congress passed the American Recovery and Reinvestment Act of 2009, commonly referred to as the "stimulus package."

annual fuel utilization efficiency (AFUE) – A measure of average fuel-use efficiency of combustion-type equipment such as furnaces and boilers over a year calculated as the ratio of annual output energy to annual input energy.

approved – Acceptable to the code official or other authority having jurisdiction. Approval is the result of investigation or by reason of accepted principles understood to fulfill the intent of specific code provisions.

ASHRAE – American Society of Heating, Refrigerating, and Air-conditioning Engineers.

ASTM International – Formerly known as the American Society for Testing and Materials; a globally recognized leader in the development and delivery of international voluntary consensus standards.

authority having jurisdiction (AHJ) – The official or entity with authority to enforce the jurisdiction's adopted codes and ordinances.

automatic – Self-acting, operating by its own mechanism.

B

basement wall – In the energy code for residential buildings, when more than 50 percent of a wall is below grade it is considered a basement wall. The measurement is the percentage of the portion of the building foundation or other exterior walls enclosing space above or below grade. For commercial prescriptive requirements, see also Section C402.2.2.

Btu (British thermal unit) – The amount of heat required to raise the temperature of a pound of water 1°F. Burning a wooden kitchen match produces about 1 Btu.

building envelope – The components of the building exterior that separate the outside environment from the inside environment.

building thermal envelope – Basement walls, exterior walls, floor, roof, any other building element part of the six-side cube that provides a boundary between conditioned space and unconditioned space.

C

coefficient of performance (COP) – Of two types, cooling and heating. COP-Cooling is a measure of performance used for refrigeration systems; COP-Heating is a measure of performance used for heat pump systems.

commercial building – In the context of the energy code, all buildings that are not residential buildings.

commissioning – A process of verifying and documenting that selected building systems or equipment perform and function as originally designed and intended.

compact fluorescent light (CFL) – A lamp of a small compact shape with a single base that uses electricity to excite mercury vapor that produces ultraviolet light, producing a visible light. A fluorescent lamp converts electrical power into useful light much more efficiently than incandescent lamps; incandescent lamp light is produced by electric resistance, such as the glow of an electric oven element.

conditioned floor area – The floor area of the space or building expressed in square feet and easily found on the floor plan in building documents. Only areas within the building that are heated or cooled count as conditioned floor area.

conditioned space – An area, room or space being heated or cooled by any equipment or appliance. In the energy code, the conditioned space may contain uninsulated ducts or an opening directly into an adjacent conditioned space.

continuous air barrier – An assembly or building material that restricts or resists airflow into or out of the building through the building thermal envelope.

continuous insulation (CI) – Insulation that is continuous across all structural members with no significant thermal breaks (thermal bridges).

cooling degree days (CDD) – The number of degrees difference each day above 65°F. If the average outside temperature is 80°F for 10 days, the cooling degree days for that time period is 150 CCD [(80°F – 65°F) 5 (15°F 3 10 days) 5 150 CDD]. The days are added together for the entire cooling season.

D

degree day (DD) – A unit of heat measurement equal to 1°F difference from a standard temperature to the average temperature of one day. The standard temperature is 65°F.

department of energy (DOE) – U.S. Department of Energy, created in 1977. The DOE is active in energy code development.

duct – A passageway usually made of sheet metal or a flexible material used to convey air throughout, into and out of the building envelope.

dynamic glazing – Fenestration products that have the ability to change their performance properties, allowing the occupant to manually control the environment by tinting (or darkening) a window or change other features or fenestration products that change their performance automatically in response to a control or environmental signal.

E

energy – The capacity for doing work. Thermal energy is expressed in Btus; electrical energy is measured in kilowatt-hours.

energy efficiency ratio (EER) – The ratio of usable output to input of energy. The output number is Btus per hour of performance to the input number of the number of watts used by the mechanical component.

Energy Rating Index (ERI) – a numerical integer value from 0 to 100 based on a linear scale between a residential building that uses no net purchased energy and one that meeds the minimum requirements of the 2006 *International Energy Conservation Code*.

enthalpy – In the case of air, the total energy of the air including the water contained in the air. For example, conditioning "dry" air requires less energy than conditioning "humid" air.

F

fenestration – Skylights, roof windows, vertical windows, opaque doors, glazed doors, and glazed block, including products with glass and nonglass glazing materials.

flue – A pipe or shaft that contains exhaust gases and carries them to the outside environment.

H

heating degree day (HDD) – The number of degrees difference each day below 65°F. If the average outside temperature is 40°F for 10 days, the heating degree days for that time period is 250 HHD [(65°F 2 40°F) 5 (25°F 3 10 days) 5 250 HDD]. The days are added together for the entire heating season.

heating, ventilation and air conditioning (HVAC) – The mechanical system that controls the air temperature and change of air within a space or building for health and comfort.

I

IBC – *International Building Code* published by the ICC.

ICC – International Code Council, the publisher of the family of I-Codes, including the IBC, IRC, and the IECC.

IECC – *International Energy Conservation Code* published by the ICC.

Illuminating Engineering Society of North America (IESNA) – The association dedicated to improving the lighted environment; founded in 1906.

infiltration – Uncontrolled flow of air into the building, whether intentional or not, through windows, doors, cracks and seams, or other areas that allow air to penetrate.

insulation – The material used as part of the building thermal envelope to slow the flow of heat through the enclosure. Insulation is made from a number of different materials, including glass or mineral fiber, cellulose, polystyrene and polyurethane foam. It can be loose filled and blown into wall cavities and attic space, or installed as batts, boards or blocks in the thermal envelope.

J

jurisdiction – The government unit that adopts codes or regulations under due legislative authority.

L

light-emitting diode (LED) – An electronic device or light source illuminated just by the movement of electrons in a semiconductor; LEDs do not get very hot.

M

manual – Capable of being operated by personal intervention.

mass wall – Masonry, concrete or similar dense building material weighing at least 35 pounds per square foot or solid wood or other less dense material weighing at least 25 pounds per square foot. These wall material types provide thermal storage benefits. Examples include adobe, filled concrete blocks, poured-in-place reinforced concrete and sawn logs.

N

National Appliance Energy Conservation Act (NAECA) – The National Appliance Energy Conservation Act of 1987 established minimum efficiency standards for many household appliances, such as air conditioners, clothes dryers, clothes washers, dishwashers, refrigerators, freezers, kitchen ranges, ovens and pool heaters.

National Fenestration Rating Council (NFRC) – A nonprofit organization formed in response to the energy crisis of the early 1970s that administers a uniform and independent rating and labeling system for the energy performance of windows, doors, skylights and attachment products.

O

outdoor air – Air introduced into the building from the outdoors and not previously circulated through conditioner equipment.

P

performance – A methodology based on objectives and desired outcomes, rather than the specifics of how to achieve it.

prescriptive – A methodology that provides specific steps or actions to result in a desired outcome.

projection factor (PF) – The horizontal depth of the overhang divided by the distance from the bottom of the fenestration to the lowest point of the bottom of the overhang.

R

recirculated air – Air returned from the inside of the building that passes through the conditioner before being supplied again to the conditioned space.

residential building – In the context of energy code, detached one- and two-family dwellings and multiple single-family dwellings (townhouses) as well as Group R-2, R-3 and R-4 buildings three stories or less in height. (For all occupancy classifications, including R-2, R-3 and R-4, refer to the IBC.)

return air – Air reintroduced into the building without passing through the conditioner.

***R*-value** – Unit of measurement for thermal resistance of materials or assemblies. The higher the *R*-value of a material or assembly, the less heat or cold passes through and the more energy efficient the insulation.

S

sealant – A mixture used to fill and seal joints where some movement between materials is expected.

seasonal energy efficiency ratio (SEER) – Unit of measurement for the efficiency of air-conditioning systems calculated by dividing the total cooling output during the normal annual cooling period usage in Btu by total electric energy input during the same period in watt-hour (Wh).

solar heat gain coefficient (SHGC) – A measure of how well a window or skylight assembly blocks the heat from sunlight, expressed as a number between 0 and 1. The lower the SHGC, the more heat is reflected by the window, resulting in greater energy efficiency in hot climates.

stack effect – The upward and downward movement of air in a building due to temperature difference. Pressure caused by the air movement can lead to air leakage into and out of the building envelope.

T

thermal bridge – An element in a building or building envelope assembly that provides less thermal resistance than the adjacent construction.

tons of cooling – Air-conditioning unit cooling capacity is often expressed as "tons of AC." 12,000 Btus of cooling equals 1 ton. An 8-ton air conditioner produces 96,000 Btus of cooling capacity.

U

UA – The thermal transmittance of a specific contact area.

U-factor – A measure of heat conduction for building elements or assemblies. It is the inverse of R-value (1/R).

V

vent – A pipe or conduit composed of factory-made components for conveying air or gasses to the atmosphere.

ventilation – The natural or mechanical process of supplying air to or removing air from a room or space.

vestibule – An intermediate space near the primary entrance doors separating the indoor conditioned air from the outside air and intended to reduce infiltration of outside air.

visible transmittance (VT) – A measure of the amount of visible light transmitted for an assembly, expressed in units between 0 and 1. The higher the VT, the more light is transmitted and the more efficient is the use of daylight.

W

watt – The unit of measurement for electrical power.

weatherstripping – The process of sealing or the material used to seal around openings such as doors and windows to retard the infiltration or exfiltration of air.

INDEX

Information in figures and tables is denoted by *f* and *t*.

ICC INTERNATIONAL CODE COUNCIL®
People Helping People Build a Safer World®

The ICC Assessment Center (formerly known as ICC Certification & Testing) provides nationally recognized credentials that demonstrate a confirmed commitment to protect public health, safety, and welfare. Raise the professionalism of your department and further your career by pursuing an ICC Certification.

ICC Certifications offer:

- Nationwide recognition
- Increased earning potential
- Career advancement
- Superior knowledge
- Validation of your expertise
- Personal and professional satisfaction

Exams are developed and maintained to the highest standards, which includes continuous peer review by national committees of experienced, practicing professionals. ICC is continually evolving exam offerings, testing options, and technology to ensure that all building and fire safety officials have access to the tools and resources needed to advance in today's fast-paced and rapidly-changing world.

Enhancing Exam Options

Effective July 2018, the Assessment Center enhanced and streamlined exam options and now offers only computer based testing (CBT) at a test site and PRONTO. We no longer offer paper/pencil exams.

Proctored Remote Online Testing Option (PRONTO)

Taking your next ICC certification exam is more convenient, more comfortable and more efficient than ever before with PRONTO.

pronto
ICC ONLINE EXAMS

PRONTO provides a convenient testing experience that is accessible 24 hours a day, 7 days a week, 365 days a year. Required hardware/software is minimal – you will need a webcam and microphone, as well as a reasonably recent operating system.

Whether testing in your office or in the comfort of your home, your ICC exam will continue to maintain its credibility while offering more convenience, allowing you to focus on achieving your professional goals. The Assessment Center continues to add exams to the PRONTO exam catalog regularly.

18-15616

Checkout all the ICC Assessment Center has to offer at iccsafe.org/certification